Silvanus Phillips Thompson

Lectures on the Electromagnet

Silvanus Phillips Thompson

Lectures on the Electromagnet

ISBN/EAN: 9783744674898

Printed in Europe, USA, Canada, Australia, Japan

Cover: Foto ©berggeist007 / pixelio.de

More available books at **www.hansebooks.com**

LECTURES ON

THE ELECTROMAGNET.

BY

SILVANUS P. THOMPSON, D.Sc., B.A., F.R.A.S.

PRINCIPAL OF, AND PROFESSOR OF PHYSICS IN THE CITY AND GUILDS
OF LONDON TECHNICAL COLLEGE, FINSBURY ; VICE PRESIDENT
OF THE PHYSICAL SOCIETY OF LONDON, ETC., ETC.

AUTHORIZED AMERICAN EDITION

NEW YORK:
THE W. J. JOHNSTON COMPANY, Ltd.
1891.

NOTICE.

This course of four Lectures on the Electromagnet was delivered in February, 1890, before the Society of Arts, London, and constituted one of the sets of "Cantor" Lectures of the Session 1889-90. This volume is reprinted with the direct sanction of the Author, who has revised the text for republication. It is the only authorized American edition.

CONTENTS.

	PAGE
LECTURE I.,	7

Introductory; Historical Sketch; Generalities Concerning Electromagnets; Typical Forms; Polarity; Uses in General; The Properties of Iron; Methods of Measuring Permeability; Traction Methods; Curves of Magnetization and Permeability; The Law of the Electromagnet; Hysteresis; Fallacies and Facts about Electromagnets.

LECTURE II., ... 80

General Principles of Design and Construction; Principles of the Magnetic Circuit.

APPENDIX TO LECTURE II., ... 161

Calculation of Excitation, Leakage, etc.; Rules for Estimating Magnetic Leakage.

LECTURE III., ... 171

Special Designs; Winding of the Copper; Windings for Constant Pressure and for Constant Current; Miscellaneous Rules about Winding; Specifications of Electromagnets; Amateur Rule about Resistance of Electromagnet and Battery; Forms of Electromagnets; Effect of Size of Coils; Effect of Position of Coils; Effect of Shape of Section; Effect of Distance between Poles; Researches of Prof. Hughes; Position and Form of Armature; Pole-Pieces on Horseshoe Magnets; Contrasts between Electromagnets and Permanent Magnets; Electromagnets for Maximum Traction; Electromagnets for Maximum Range of Attraction; Electromagnets of Minimum Weight; A Useful Guiding Principle; Electromagnets for Use with Alternating Currents; Electromagnets for Quickest Action; Connecting Coils for Quickest Action; Battery Grouping for Quickest Action; Time-Constant of Electromagnets; Short Cores vs. Long Cores.

LECTURE IV., ... 222

Electromagnetic Mechanism; The Coil-and-Plunger; Effect of Using Coned Plunger; Other Modes of Extending Range of Action; Modifications of the Coil-and-Plunger; Differential Coil-and-Plunger; Coil-and-Plunger Coil; Intermediate Forms; Action of Magnetic Field on Small Iron Sphere; Sectioned Coils with Plunger; Winding of Tubular Coils and Electromagnets; Extension of Range by Oblique Approach; Polarized Mechanism; Uses of Permanent Magnets; Electromagnetic Mechanism; Moving Coil in Permanent Magnetic Field; Magnetic Adherence; Repulsion Mechanism; Electromagnetic Vibrators; Indicator Movements; The Study of Electromagnetic Mechanism; Suppression of Sparking; Conclusion.

LIST OF ILLUSTRATIONS.

	PAGE
Sturgeon's First Electromagnet,	18
Sturgeon's Straight-Bar Electromagnet,	19
Sturgeon's Lecture-Table Electromagnet,	25
Henry's Electromagnet,	35
Henry's Experimental Electromagnet,	36
Joule's Electromagnet,	41
Joule's Cylindrical Electromagnet,	45
Roberts' Electromagnet,	46
Joule's Zigzag Electromagnet,	46
Typical Two-Pole Electromagnet,	50
Iron-Clad Electromagnet,	50
Diagram Illustrating Relation of Magnetizing Circuit and Resulting Magnetic Force,	51
Curves of Magnetization of Different Magnetic Materials,	57
Ring Method of Measuring Permeability (Rowland's Arrangement),	60
Bosanquet's Data of Magnetic Properties of Iron and Steel Rings,	62
Hopkinson's Divided Bar Method of Measuring Magnetic Permeability,	64
Curves of Magnetization of Iron,	66
The Permeameter,	70
Curves of Permeability,	73
Curves of Hysteresis,	75
Bosanquet's Verification of the Law of Traction,	90
Stumpy Electromagnet,	97
Experiment on Rounding Ends,	105
Experiment of Detaching Armature,	105
Lines of Force Running through Bar Magnet,	107
Apparatus to Illustrate the Law of Inverse Squares,	113
Deflection of Needle Caused by Bar Magnet Broadside on,	115
Closed Magnetic Circuit,	116
Divided Magnetic Circuit,	117
Electromagnet with Armature in Contact,	119
Electromagnet with Air-Gaps One Millimetre Wide,	119
Electromagnet with Air-Gaps Several Millimetres Wide,	121
Electromagnet without Armature,	121

LIST OF ILLUSTRATIONS.

	PAGE
Contrasted Effect of Flat and Pointed Poles,	127
Dub's Experiments with Pole-Pieces,	129
Dub's Deflection Experiment,	130
Deflecting a Steel Magnet Having Bifilar Suspension—Pole-Piece on Near End,	131
Deflecting Steel Magnet—Pole-Piece on Distant End,	131
Experiment with Tubular Core and Iron Ring,	136
Exploring Polar Distribution with Small Iron Ball,	137
Iron Ball Attracted to Edge of Polar Face,	139
Experiment on Leakage of Electromagnet,	140
Curves of Magnetization Plotted from Preceding,	143
Curves of Flow of Magnetic Lines in Air from One Cylindrical Pole to Another,	146
Diagram of Leakage Reluctances,	148
Von Feilitzsch's Curves of Magnetization of Rods of Various Diameters,	152
Ewing's Curves for Effect of Joints,	157
Von Feilitzsch's Curves of Magnetization of Tubes,	159
Club-Footed Electromagnet,	189
Hughes' Electromagnet,	195
Experiment with Permanent Magnet,	200
Electromagnetic Pop-Gun,	205
Curves of Rise of Currents,	211
Curves of Rise of Current with Different Groupings of Battery,	216
Electromagnets of Relay and their Effects,	219
Hjörth's Electromagnetic Mechanism,	224
Action of Single Coil on Point Pole on Axis,	230
Action along Axis of Single Coil,	230
Action of Tubular Coil,	232
Diagram of Force and Work of Coil-and-Plunger,	235
Von Feilitzsch's Experiment on Plungers of Iron and Steel,	244
Bruger's Experiments on Coils and Plungers,	245
Bruger's Experiments, Using Currents of Various Strengths,	245
Plunger Electromagnet of Stevens and Hardy,	252
Electromagnet of Brush Arc Lamp,	253
Ayrton and Perry's Tubular Iron-Clad Electromagnet,	254
Froment's Equalizer with Stanhope Lever,	261
Davy's Mode of Controlling Armature by Spring,	261
Robert Houdin's Equalizer,	262
Mechanism of Duboscq's Arc Lamp,	263
Nicklès' Magnetic Friction Gear,	271
Forbes' Electromagnet,	272
Electromagnetic Mechanism Working by Repulsion,	273
Repulsion between Two Parallel Cores,	273

THE
ELECTROMAGNET.

LECTURE I.

INTRODUCTORY.

AMONG the great inventions which have originated in the lecture-room in which we are met are two of special interest to electricians—the application of gutta-percha to the purpose of submarine telegraph cables, and the electromagnet. This latter invention was first publicly described, from the very platform on which I stand, on May 23, 1825, by William Sturgeon, whose paper is to be found in the forty-third volume of the *Transactions of the Society of Arts*. For this invention we may rightfully claim the very highest place. Electrical engineering, the latest and most vigorous offshoot of applied science, embraces many branches. The dynamo for generating electric currents, the motor for transforming their energy back into work, the arc lamp, the electric bell, the telephone, the recent electromagnetic machinery for coal-mining, for the separation of ore, and many other electro-mechanical contrivances, come within the purview of the electrical engineer. In every one of these, and in many more of the useful applications of electricity, the central organ is an electromagnet. By means of this simple and familiar contrivance—an iron core surrounded by a copper-wire coil—mechanical actions are produced at will, at a distance, under control, by the agency of electric currents. These mechanical actions are known to vary with the mass, form, and quality of the iron core, the quantity and disposition of the copper wire wound upon it, the quantity of electric current circulating around it, the form, quality, and distance of the iron armature upon which it acts. But the laws which govern the mechanical action in relation to these various matters are by no means well known, and, indeed, several of them have long been a matter of dispute. Gradually, however, that which has been vague and indeterminate becomes clear and precise. The laws of the steady circulation of electric currents, at one time altogether obscure, were cleared up by the discovery of the famous law of Ohm. Their extension to the case of rapidly interrupted currents, such as are used in telegraphic working, was discovered by Helmholtz; while to Maxwell is due their future extension to alternating, or, as they are sometimes called, undulatory currents. All this was purely electric work. But the law of the electromagnet was still undiscovered; the magnetic part of the problem was still buried in obscurity. The only exact reasoning about magnetism dealt with problems of another kind; it was couched in language of a misleading character; for the practical problems connected with the electromagnet it was worse than useless. The doctrine of two magnetic fluids distributed over the end surfaces of magnets, under the sanction of the great names of Coulomb, of Poisson, and of Laplace, had unfortunately become recognized as an accepted part of science along with the law of inverse squares. How greatly the progress of electromagnetic science has been impeded and retarded by the weight of these great names it is impossible now to gauge. We now know that for all purposes, save only those whose value lies in the domain of abstract mathematics, the doctrine of the two magnetic fluids is false and misleading. We know that magnetism, so far from residing on the end or surface of the magnet, is a property resident throughout the mass; that the internal, not the external, magnetization is the important fact to be considered; that the so-called free magnetism on the surface is, as it were, an accidental phenomenon; that the magnet is really most highly magnetized at those parts where there is least surface magnetization; finally, that the doctrine of surface distribution of fluids is absolutely incompetent to afford a basis of calculation such as is required by the electrical engineer. He requires rules to enable him not only to predict the lifting power of a given electromagnet, but also to guide him in designing and constructing electromagnets of special forms suitable for the various cases that arise in his practice. He wants in one place a strong electromagnet to hold on to its armature like a limpet to its native rock; in another case he desires a magnet having a very long range of attraction, and wants a rule to guide him to the best design; in another he wants a special form having the most rapid action attainable; in yet another he must sacrifice everything else to attain maximum action with minimum weight. Toward the solution of such practical problems as these the old theory of magnetism offered not the slightest aid. Its array of mathematical symbols was a mockery. It was as though an engineer asking for rules to enable him to design the cylinder and piston of an engine were confronted with recipes how to estimate the cost of painting it.

Gradually, however, new light dawned. It became customary, in spite of the mathematicians, to regard the magnetism of a magnet as something that traverses or circulates around a definite path, flowing more freely through such substances as iron than through other relatively non-magnetic materials. Analogies between the flow of electricity in an electrically conducting circuit, and the passage of magnetic lines of force through circuits possessing magnetic conductivity, forced themselves upon the minds of experimenters, and compelled a mode of thought quite other than that previously accepted. So far back as 1821, Cumming[1] experimented on magnetic conductivity. The idea of a magnetic circuit was more or less familiar to Ritchie,[2] Sturgeon,[3] Dove,[4] Dub,[5] and De La Rive,[6] the last-named of whom

---

[1] *Camb. Phil. Trans.*, Apr. 2, 1821.

[2] *Phil. Mag.*, series iii., vol. iii., p. 122.

[3] *Ann. of Electr.*, xii., p. 217.

[4] *Pogg. Ann.*, xxix., p. 462, 1833. See also *Pogg. Ann.*, xliii., p. 517, 1838.

[5] Dub, "Elektromagnetismus" (ed. 1861), p. 401 ; and *Pogg. Ann.*, xc., p. 440, 1853.

[6] De La Rive, "Treatise on Electricity" (Walker's translation), vol. i., p. 292.

explicitly uses the phrase, "a closed magnetic circuit."
Joule[7] found the maximum power of an electromagnet
to be proportional to "the least sectional area of the entire magnetic circuit," and he considered the resistance
to induction as proportional to the length of the magnetic circuit. Indeed, there are to be found scattered
in Joule's writings on the subject of magnetism, some
five or six sentences, which, if collected together, constitute a very full statement of the whole matter. Faraday[8] considered that he had proved that each magnetic
line of force constitutes a closed curve; that the path of
these closed curves depended on the magnetic conductivity of the masses disposed in proximity; that the
lines of magnetic force were strictly analogous to the
lines of electric flow in an electric circuit. He spoke of
a magnet surrounded by air being like unto a voltaic
battery immersed in water or other electrolyte. He
even saw the existence of a power, analogous to that of
electromotive force in electric circuits, though the name,
"magneto-motive force," is of more recent origin. The
notion of magnetic conductivity is to be found in Maxwell's great treatise (vol. ii., p. 51), but is only briefly
mentioned. Rowland,[9] in 1873, expressly adopted the
reasoning and language of Faraday's method in the working out of some new results on magnetic permeability,
and pointed out that the flow of magnetic lines of force

---

[7] *Ann. of Electr.*, iv., 59, 1839; v., 195, 1841; and "Scientific Papers," pp. 8, 31, 35, 36.

[8] "Experimental Researches," vol. iii., art. 3117, 3228, 3230, 3260, 3271, 3276, 3294, and 3361.

[9] *Phil. Mag.*, series iv., vol. xlvi., Aug., 1873, "On Magnetic Permeability and the Maximum of Magnetism of Iron, Steel, and Nickel."

through a bar could be subjected to exact calculation; the elementary law, he says, "is similar to the law of Ohm." According to Rowland, the "magnetizing force of helix" was to be divided by the "resistance to the lines of force;" a calculation for magnetic circuits which every electrician will recognize as precisely Ohm's law for electric circuits. He applied the calculations to determine the permeability of certain specimens of iron, steel, and nickel. In 1882,[10] and again in 1883, Mr. R. H. M. Bosanquet[11] brought out at greater length a similar argument, employing the extremely apt term "magneto-motive force" to connote the force tending to drive the magnetic lines of induction through the "magnetic resistance," or, as it will frequently be called in these lectures, the magnetic "reluctance," of the circuit. In these papers the calculations are reduced to a system, and deal not only with the specific properties of iron, but with problems arising out of the shape of the iron. Bosanquet shows how to calculate the several resistances (or reluctances) of the separate parts of the circuit, and then add them together to obtain the total resistance (or reluctance) of the magnetic circuit.

Prior to this, however, the principle of the magnetic circuit had been seized upon by Lord Elphinstone and Mr. Vincent, who proposed to apply it in the construction of dynamo-electric machines. On two occasions[12]

---

[10] *Proc. Roy. Soc.*, xxxiv., p. 445, Dec., 1882.

[11] *Phil. Mag.*, series v., vol. xv., p. 205, Mar., 1883, "On Magneto-Motive Force." Also *ib.*, vol. xix., Feb., 1885, and *Proc. Roy. Soc.*, No. 228, 1883. See also *The Electrician* (London), xiv., p. 291, Feb. 14, 1885.

[12] *Proc. Roy. Soc.*, xxix., p. 292, 1879, and xxx., p. 287, 1880. See *Electrical Review* (London), viii., p. 134, 1880.

they communicated to the Royal Society the results of experiments to show that the same exciting current would evoke a larger amount of magnetism in a given iron structure, if that iron structure formed a closed magnetic circuit than if it were otherwise disposed.

In recent years the notion of the magnetic circuit has been vigorously taken up by the designers of dynamo machines, who, indeed, base the calculation of their designs upon this all-important principle. Having this, they need no laws of inverse squares of distances, no magnetic moments, none of the elaborate expressions for surface distribution of magnetism, none of the ancient paraphernalia of the last century. The simple law of the magnetic circuit and a knowledge of the properties of iron are practically all they need. About four years ago, much was done by Mr. Gisbert Kapp [13] and by Drs. J. and E. Hopkinson [14] in the application of these considerations to the design of dynamo machines, which previously had been a matter of empirical practice. To this end the formulæ of Professor Forbes [15] for calculating magnetic leakage, and the researches of Professors Ayrton and Perry [16] on magnetic shunts, contributed a not unimportant share. As the result of the advances made at that time, the subject of dynamo design was reduced to an exact science.

It is the aim and object of the present course of lec-

---

[13] *The Electrician* (London), vols. xiv., xv., and xvi., 1885-86; also *Proc. Inst. Civil Engineers*, lxxxiii., 1885-86; and *Jour. Soc. Telegr. Engineers*, xv., 524, 1886.

[14] *Phil. Trans.*, 1886, pt. i., p. 331; and *The Electrician* (London), xviii., pp. 39, 63, 86, 1886.

[15] *Jour. Soc. Telegr. Engineers*, xv., 555, 1886.

[16] *Jour. Soc. Telegr. Engineers*, xv., 530, 1886.

tures to show how the same considerations which have been applied with such great success to the subject of the design of dynamo-electric machines may be applied to the study of the electromagnet. The theory and practice of the design and construction of electromagnets will thus be placed, once for all, upon a rational basis. Definite rules will be laid down for the guidance of the constructor, directing him as to the proper dimensions and form of iron to be chosen, and as to the proper size and amount of copper wire to be wound upon it in order to produce any desired result.

First, however, a historical account of the invention will be given, followed by a number of general considerations respecting the uses and forms of electromagnets. These will be followed by a discussion of the magnetic properties of iron and steel and other materials; some account being added of the methods used for determining the magnetic permeability of various brands of iron at different degrees of saturation. Tabular information is given as to the results found by different observers. In connection with the magnetic properties of iron, the phenomenon of magnetic hysteresis is also described and discussed. The principle of the magnetic circuit is then discussed with numerical examples, and a number of experimental data respecting the performance of electromagnets are adduced, in particular those bearing upon the tractive power of electromagnets. The law of traction between an electromagnet and its armature is then laid down, followed by the rules for predetermining the iron cores and copper coils required to give any prescribed tractive force.

Then comes the extension of the calculation of the magnetic circuit to those cases where there is an air-gap between the poles of the magnet and the armature, and where, in consequence, there is leakage of the magnetic lines from pole to pole. The rules for calculating the winding of the copper coils are stated, and the limiting relation between the magnetizing power of the coil and the heating effect of the current in it is explained. After this comes a detailed discussion of the special varieties of form that must be given to electromagnets in order to adapt them to special services. Those which are designed for maximum traction, for quickest action, for longest range, for greatest economy when used in continuous daily service, for working in series with constant current, for use in parallel at constant pressure, and those for use with alternate currents are separately considered.

Lastly, some account is given of the various forms of electromagnetic mechanism which have arisen in connection with the invention of the electromagnet. The plunger and coil is specially considered as constituting a species of electromagnet adapted for a long range of motion. Modes of mechanically securing long range for electromagnets and of equalizing their pull over the range of motion of the armature are also described. The analogies between sundry electro-mechanical movements and the corresponding pieces of ordinary mechanism are traced out. The course is concluded by a consideration of the various modes of preventing or minimizing the sparks which occur in the circuits in which electromagnets are used.

HISTORICAL SKETCH.

The effect which an electric current, flowing in a wire, can exercise upon a neighboring compass needle was discovered by Oersted in 1820.[17] This first announcement of the possession of magnetic properties by an electric current was followed speedily by the researches of Ampère,[18] Arago,[19] Davy,[20] and by the devices of several other experimenters, including De La Rive's[21] floating battery and coil; Schweigger's[22] multiplier, Cumming's[23] galvanometer, Faraday's[24] apparatus for rotation of a permanent magnet, Marsh's[25] vibrating pendulum, and Barlow's[26] rotating star-wheel. But it was not until 1825 that the electromagnet was invented. Davy had, indeed, in 1821, surrounded with temporary coils of wire the steel needles upon which he was experimenting, and had shown that the flow of electricity around the coil could confer magnetic power upon the steel needles. But from this experiment it was a grand step forward to the discovery that a core of soft iron, surrounded by its own appropriate coil of copper, could be made to act not only as a powerful magnet, but as a magnet whose power could be turned on or off at will, could be aug-

---

[17] See Thomson's *Annals of Philosophy*, Oct., 1820.
[18] *Ann. de Chim. et de Physique*, xv., 59 and 170, 1820.
[19] *Ib.*, xv., 93, 1820.
[20] *Phil. Trans.*, 1821.
[21] "Bibliothéque Universelle," Mar., 1821.
[22] *Ib.*   [23] *Camb. Phil. Trans.*, 1821.
[24] *Quarterly Journal of Science*, Sept., 1821.
[25] Barlow's "Magnetic Attractions," second edition, 1823.
[26] *Ib.*

mented to any desired degree, and could be set into action and controlled from a practically unlimited distance.

The electromagnet, in the form which can first claim recognition for these qualities, was devised by William Sturgeon,[27] and is described by him in the paper which he contributed to the proceedings of the Society of Arts in 1825, accompanying a set of improved apparatus for electromagnetic experiments.[28] The Society of Arts rewarded Sturgeon's labors by awarding him the silver medal of the society and a premium of 30 guineas. Among this set of apparatus are two electromagnets,

---

[27] William Sturgeon, the inventor of the electromagnet, was born at Whittington, in Lancashire, in 1783. Apprenticed as a boy to the trade of a shoemaker, at the age of 19 he joined the Westmoreland militia, and two years later enlisted into the Royal Artillery, thus gaining the chance of learning something of science, and having leisure in which to pursue his absorbing passion for chemical and physical experiments. He was 42 years of age when he made his great, though at the time unrecognized, invention. At the date of his researches in electromagnetism he was resident at 8 Artillery place, Woolwich, at which place he was the associate of Marsh and was intimate with Barlow, Christie, and Gregory, who interested themselves in his work. In 1835 he presented a paper to the Royal Society containing descriptions, *inter alia*, of a magneto-electric machine with longitudinally wound armature, and with a commutator consisting of half discs of metal. For some reason this paper was not admitted to the *Philosophical Transactions;* he afterward printed it in full, without alteration, in his volume of "Scientific Researches," published by subscription in 1850. From 1836 to 1843 he conducted the *Annals of Electricity*. He had now removed to Manchester, where he lectured on electricity at the Royal Victoria Gallery. He died at Prestwick, near Manchester, in 1850. There is a tablet to his memory in the church at Kirkby Lonsdale, from which town the village of Whittington is distant about two miles. A portrait of Sturgeon in oils, and said to be an excellent likeness, is believed still to be in existence; but all inquiries as to its whereabouts have proved unavailing. At the present moment, so far as I am aware, the scientific world is absolutely without a portrait of the inventor of the electromagnet.

[28] *Trans. Society of Arts*, 1825, xliii., p. 38

one of horseshoe shape (Figs. 1 and 2) and one a straight bar (Fig. 3). It will be seen that the former figures present an electromagnet consisting of a bent iron rod about one foot long and a half inch in diameter, varnished over and then coiled with a single left-handed spiral of stout uncovered copper wire of 18 turns. This

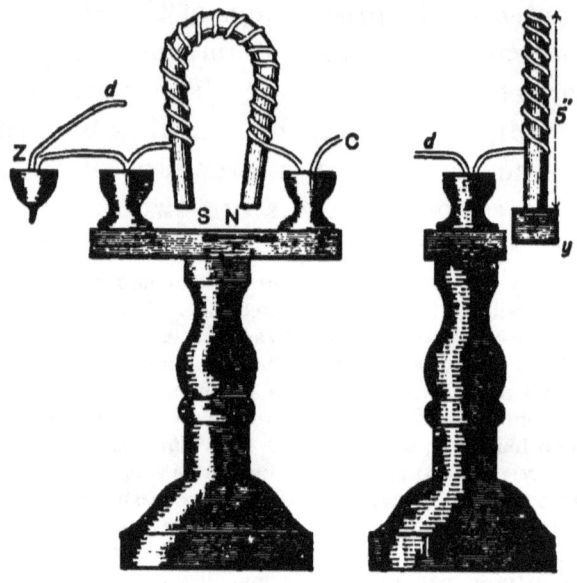

FIGS. 1 AND 2.—STURGEON'S FIRST ELECTROMAGNET.

coil was found appropriate to the particular battery which Sturgeon preferred, namely, a single cell containing a spirally enrolled pair of zinc and copper plates of large area (about 130 square inches) immersed in acid; which cell, having small internal resistance, would yield a large quantity of current when connected to a circuit of small resistance. The ends of the copper wire were brought out sideways and bent down so as to dip in two deep connecting cups marked *Z* and *C*, fixed upon a wooden stand. These cups, which were of wood, served as supports to hold up the electromagnet, and having mercury in them served also to make good electrical connection. In Fig. 2 the magnet is seen sideways, supporting a bar of iron, *y*. The circuit was completed to the battery through a connecting wire, *d*, which could be lifted out of the cup, *Z*, so breaking circuit when desired, and allowing the weight to drop. Sturgeon added in his explanatory remarks that the poles, *N* and *S*, of the magnet will be reversed if you wrap the copper wire about the rod as a right-handed screw, instead of a left-handed one, or, more simply, by reversing the connections with the battery, by causing the wire that dips into the *Z* cup to dip into the *C* cup, and *vice versa*. This electromagnet was capable of supporting nine pounds when thus excited.

Fig. 3.—Sturgeon's Straight-Bar Electromagnet.

Fig. 3 shows another arrangement to fit on the same stand. This arrangement communicates magnetism to hardened steel bars as soon as they are put in, and renders soft iron within it magnetic during the time of action; it only differs from Figs. 1 and 2 in being straight, and thereby allows the steel or iron bars to slide in and out.

For this piece of apparatus and other adjuncts accompanying it, all of which are described in the Society's *Transactions* for 1825, Sturgeon, as already stated, was awarded the society's silver medal and a premium of 30 guineas. The apparatus was deposited in the museum of the society, which therefore might be supposed to be the proud possessor of the first electromagnet ever constructed. Alas! for the vanity of human affairs, the society's museum of apparatus has long been dispersed, this priceless relic having been either made over to the now defunct Patent-office Museum or otherwise lost sight of.

Sturgeon's first electromagnet, the core of which weighed about seven ounces, was able to sustain a load of nine pounds, or about 20 times its own weight. At the time it was considered a truly remarkable performance. Its single layer of stout copper wire was well adapted to the battery employed, a single cell of Sturgeon's own particular construction having a surface of 130 square inches, and therefore of small internal resistance. Subsequently, in the hands of Joule, the same electromagnet sustained a load of 50 pounds, or about 114 times its own weight. Writing in 1832 about his apparatus of 1825, Sturgeon used the following magniloquent language:

"When first I showed that the magnetic energies of a galvanic conducting wire are more conspicuously exhibited by exercising them on soft iron than on hard steel, my experiments were limited to small masses—generally to a few inches of rod iron about half an inch in diameter. Some of those pieces were employed while straight, and others were

LECTURES ON THE ELECTROMAGNET.

bent into the form of a horseshoe magnet, each piece being compassed by a spiral conductor of copper wire. The magnetic energies developed by these simple arrangements are of a very distinguished and exalted character, as is conspicuously manifested by the suspension of a considerable weight at the poles during the period of excitation by the electric influence.

"An unparalleled transiliency of magnetic action is also displayed in soft iron by an instantaneous transition from a state of total inactivity to that of vigorous polarity, and also by a simultaneous reciprocity of polarity in the extremities of the bar—versatilities in this branch of physics for the display of which soft iron is pre-eminently qualified, and which, by the agency of electricity, become demonstrable with the celerity of thought, and illustrated by experiments the most splendid in magnetics. It is, moreover, abundantly manifested by ample experiments, that galvanic electricity exercises a superlative degree of excitation on the latent magnetism of soft iron, and calls for its recondite powers with astonishing promptitude, to an intensity of action far surpassing anything which can be accomplished by any known application of the most vigorous permanent magnet, or by any other mode of experimenting hitherto discovered. It has been observed, however, by experimenting on different pieces selected from various sources, that, notwithstanding the greatest care be observed in preparing them of a uniform figure and dimensions, there appears a considerable difference in the susceptibility which they individually possess of developing the magnet powers, much of which depends upon the manner of treatment at the forge, as well as upon the natural character of the iron itself."[29]

---

[29] "I have made a number of experiments on small pieces, from the results of which it appears that much hammering is highly detrimental to the development of magnetism in soft iron, whether the exciting cause be galvanic or any other. And although good annealing is always essential and facilitates to a considerable extent the display of polarity, that process is

"The superlative intensity of electromagnets, and the facility and promptitude with which their energies can be brought into play, are qualifications admirably adapted for their introduction into a variety of arrangements in which powerful magnets so essentially operate and perform a distinguished part in the production of electromagnetic rotations; while the versatilities of polarity of which they are susceptible are eminently calculated to give a pleasing diversity in the exhibition of that highly interesting class of phenomena, and lead to the production of others inimitable by any other means."[30]

Sturgeon's further work during the next three years is best described in his own words:

"It does not appear that any very extensive experiments were attempted to improve the lifting power of electromagnets, from the time that my experiments were published in the *Transactions of the Society of Arts*, etc., for 1825, till the latter part of 1828. Mr. Watkins, philosophical instrument maker, Charing Cross, had, however, made them of much larger size than any which I had employed, but I am not aware to what extent he pursued the experiment.

"In the year 1828, Professor Moll, of Utrecht, being on a visit to London, purchased of Mr. Watkins an electromagnet weighing about five pounds—at that time, I believe, the largest which had been made. It was of round iron, about one inch in diameter, and furnished with a single copper wire twisted round it 83 times. When this magnet was excited by a large galvanic surface, it supported about 75 pounds. Professor Moll afterward prepared another electro-

---

very far from restoring to the iron that degree of susceptibility which it frequently loses by the operation of the hammer. Cylindric rod iron of small dimensions may very easily be bent into the required form, without any hammering whatever; and I have found that small electromagnets made in this way display the magnetic powers in a very exalted degree."

[30] Sturgeon's "Scientific Researches," p. 113.

magnet, which, when bent, was 12½ inches high, 2¼ inches in diameter, and weighed about 26 pounds, prepared, like the former, with a single spiral conducting wire. With an acting galvanic surface of 11 square feet, this magnet would support 154 pounds, but would not lift an anvil which weighed 200 pounds.

"The largest electromagnet which I have yet [1832] exhibited in my lectures weighs about 16 pounds. It is formed of a small bar of soft iron, 1¼ inches across each side; the cross-piece which joins the poles is from the same rod of iron, and about 3¾ inches long. Twenty separate strands of copper wire, each strand about 50 feet in length, are coiled around the iron, one above another, from pole to pole, and separated from each other by intervening cases of silk; the first coil is only the thickness of one ply of silk from the iron; the twentieth, or outermost, about half an inch from it. By this means the wires are completely insulated from each other without the trouble of covering them with thread or varnish. The ends of wire project about two feet for the convenience of connection. With one of my small cylindrical batteries, exposing about 150 square inches of total surface, this electromagnet supports 400 pounds. I have tried it with a larger battery, but its energies do not seem to be so materially exalted as might have been expected by increasing the extent of galvanic surface. Much depends upon a proper acid solution; good nitric or nitrous acid, with about six or eight times its quantity of water, answers very well. With a new battery of the above dimensions and a strong solution of salt and water, at a temperature of 190 degrees Fahr., the electromagnet supported between 70 and 80 pounds when the first 17 coils only were in the circuit. With the three exterior coils alone in the circuit, it would just support the lifter or cross-piece. When the temperature of the solution was between 40 and 50 degrees, the magnetic force excited was comparatively very feeble. With the innermost coil alone and a strong acid solution this electromagnet supports about 100 pounds; with the four outermost wires about 250 pounds. It improves in power with every additional coil until about the twelfth, but not perceptibly any further; therefore the remaining eight coils appear to be useless, although the last three, independently of the innermost 17, and at the distance of half an inch from the iron, produce in it a lifting power of 75 pounds.

"Mr. Marsh has fitted up a bar of iron much larger than mine with a similar distribution of the conducting wires to that devised and so successfully employed by Professor Henry. Mr. Marsh's electromagnet will support about 500 pounds when excited by a galvanic battery similar to mine. These two, I believe, are the most powerful electromagnets yet produced in this country.

"A small electromagnet, which I also employ on the lecture table, and the manner of its suspension, is represented by Fig. 3, Plate VI. The magnet is of cylindric rod iron and weighs four ounces; its poles are about a quarter of an inch asunder. It is furnished with six coils of wire in the same manner as the large electromagnet before described, and will support upward of 50 pounds.

"I find a triangular gin very convenient for the suspension of the magnet in these experiments. A stage of thin board, supporting two wooden dishes, is fastened, at a proper height, to two of the legs of the gin. Mercury is placed in these vessels, and the dependent amalgamated extremities of the conducting wires dip into it—one into each portion.

"The vessels are sufficiently wide to admit of considerable motion of the wires in the mercury without interrupting the contact, which is sometimes occasioned by the swinging of the magnet and attached weight. The circuit is completed by other wires, which connect the battery with these two portions of mercury. When the weight is supported as in the figure, if an interruption be made by removing either of the connecting wires, the weight instantaneously drops on the table. The large magnet I suspend in the same way on a larger gin; the weights which it supports are placed one after another on a square board, suspended by means of a cord at each corner from a hook in the crosspiece, which joins the poles of the magnet.

"With a new battery and a solution of salt and water, at

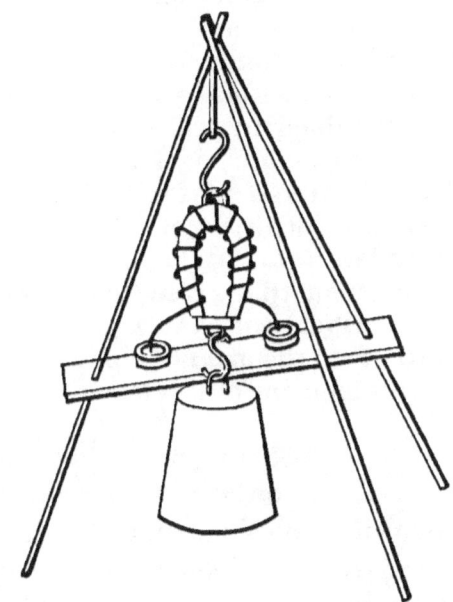

Fig. 4.—Sturgeon's Lecture-Table Electromagnet.

a temperature of 190 degrees Fahr., the small electromagnet, Fig. 3, Plate VI., supports 16 pounds." (See Fig. 4.)

In 1840, after Sturgeon had removed to Manchester, where he assumed the management of the "Victoria Gallery of Practical Science," he continued his work, and in the seventh memoir in his series of researches he wrote as follows:

"The electromagnet belonging to this institution is made of a cylindrical bar of soft iron, bent into the form of a horseshoe magnet, having the two branches parallel to each other and at the distance of 4.5 inches. The diameter of the iron is 2.75 inches; it is 18 inches long when bent. It is surrounded by 14 coils of copper wire, seven on each branch. The wire which constitutes the coils is one-twelfth of an inch in diameter, and in each coil there are about 70 feet of wire. They are united in the usual way with branch wires, for the purpose of conducting the currents from the battery. The magnet was made by Mr. Nesbit. . . . The greatest weight sustained by the magnet in these experiments is 12¾ hundred-weight, or 1,386 pounds, which was accomplished by 16 pairs of plates, in four groups of four pairs in series each. The lifting power by 19 pairs in series was considerably less than by 10 pairs in series; and but very little greater than that given by one cell or one pair only. This is somewhat remarkable, and shows how easily we may be led to waste the magnetic powers of batteries by an injudicious arrangement of its elements."[31]

At the date of Sturgeon's work the laws governing the flow of electric currents in wires were still obscure. Ohm's epoch-making enunciation of the law of the electric circuit appeared in *Poggendorff's Annalen* in the very year of Sturgeon's discovery, 1825, though his complete book appeared only in 1827, and his work, translated by Dr. Francis into English, only appeared (in Taylor's "Scientific Memoirs," vol. ii.) in 1841. Without the guidance of Ohm's law it was not strange that even the most able experimenters should not understand the relations between battery and circuit which would give them the best effects. These had to be

---

[31] Sturgeon's "Scientific Researches," p. 188.

found by the painful method of trial and failure. Preeminent among those who tried was Prof. Joseph Henry, then of the Albany Institute in New York, later of Princeton, N. J., who succeeded in effecting an important improvement. In 1828, led on by a study of the "multiplier" (or galvanometer), he proposed to apply to electromagnetic apparatus the device of winding them with a spiral coil of wire "closely turned on itself," the wire being of copper from one-fortieth to one-twenty-fifth of an inch in diameter, covered with silk. In 1831 he thus describes[32] the results of his experiments:

"A round piece of iron, about one-quarter of an inch in diameter, was bent into the usual form of a horseshoe, and instead of loosely coiling around it a few feet of wire, as is usually described, it was tightly wound with 35 feet of wire covered with silk, so as to form about 400 turns; a pair of small galvanic plates, which could be dipped into a tumbler of diluted acid, was soldered to the ends of the wire and the whole mounted on a stand. With these small plates the horseshoe became much more powerfully magnetic than another of the same size, and wound in the same manner, by the application of a battery composed of 28 plates of copper and zinc, each eight inches square. Another convenient form of this apparatus was contrived by winding a straight bar of iron nine inches long with 35 feet of wire and supporting it horizontally on a small cup of copper containing a cylinder of zinc; when this cup, which served the double purpose of a stand and the galvanic element, was filled with dilute acid the bar became a portable electromagnet. These articles were exhibited to the institute in March, 1829. The idea afterward occurred to me that a

---

[32] Silliman's *American Journal of Science*, Jan., 1831, xix., p. 400.

sufficient quantity of galvanism was furnished by the two small plates to develop, by means of the coil, a much greater magnetic power in a larger piece of iron. To test this, a cylindrical bar of iron, half an inch in diameter and about 10 inches long, was bent into the shape of a horseshoe, and wound with 30 feet of wire; with a pair of plates containing only $2\frac{1}{2}$ square inches of zinc it lifted 15 pounds avoirdupois. At the same time a very material improvement in the formation of the coil suggested itself to me on reading a more detailed account of Professor Schweigger's galvanometer, and which was also tested with complete success upon the same horseshoe; it consisted in using several strands of wire, each covered with silk, instead of one. Agreeably to this construction a second wire, of the same length as the first, was wound over it, and the ends soldered to the zinc and copper in such a manner that the galvanic current might circulate in the same direction in both, or in other words that the two wires might act as one; the effect by this addition was doubled, as the horseshoe, with the same plates before used, now supported 28 pounds.

"With a pair of plates four inches by six inches it lifted 39 pounds, or more than 50 times its own weight.

"These experiments conclusively proved that a great development of magnetism could be effected by a very small galvanic element, and also that the power of the coil was materially increased by multiplying the number of wires without increasing the number of each."[33]

Not content with these results, Professor Henry pushed forward on the line he had thus struck out. He was keenly desirous to ascertain how large a magnetic force he could produce when using only currents of such a degree of smallness as could be transmitted through the comparatively thin copper wires, such as

---

[33] "Scientific Writings of Joseph Henry," p. 39.

bell-hangers use. During the year 1830 he made great progress in this direction, as the following extracts show:

"In order to determine to what extent the coil could be applied in developing magnetism in soft iron, and also to ascertain, if possible, the most proper length of the wires to be used, a series of experiments was instituted jointly by Dr. Philip Ten Eyck and myself. For this purpose 1,060 feet (a little more than one-fifth of a mile) of copper wire of the kind called bell wire, .045 of an inch in diameter, were stretched several times across the large room of the Academy.

"*Experiment* 1.—A galvanic current from a single pair of plates of copper and zinc two inches square was passed through the whole length of the wire, and the effect on a galvanometer noted. From the mean of several observations, the deflection of the needle was 15 degrees.

"*Experiment* 2.—A current from the same plates was passed through half the above length, or 530 feet of wire; the deflection in this instance was 21 degrees.

"By a reference to a trigonometrical table, it will be seen that the natural tangents of 15 degrees and 21 degrees are very nearly in the ratio of the square roots of 1 and 2, or of the relative lengths of the wires in these two experiments.

"The length of the wire forming the galvanometer may be neglected, as it was only 8 feet long.

"*Experiment* 3.—The galvanometer was now removed, and the whole length of the wire attached to the ends of the wire of a small soft iron horseshoe, a quarter of an inch in diameter, and wound with about eight feet of copper wire with a galvanic current from the plates used in experiments 1 and 2. The magnetism was scarcely observable in the horseshoe.

"*Experiment* 4.—The small plates were removed and a battery composed of a piece of zinc plate four inches by seven inches, surrounded with copper, was substituted.

When this was attached immediately to the ends of the eight feet of wire wound round the horseshoe, the weight lifted was 4¼ pounds; when the current was passed through the whole length of wire (1,060 feet) it lifted about half an ounce.

"*Experiment 5.*—The current was passed through half the length of wire (530 feet) with the same battery; it then lifted two ounces.

"*Experiment 6.*—Two wires of the same length as in the last experiment were used, so as to form two strands from the zinc and copper of the battery; in this case the weight lifted was four ounces.

"*Experiment 7.*—The whole length of the wire was attached to a small trough on Mr. Cruickshanks' plan, containing 25 double plates, and presenting exactly the same extent of zinc surface to the action of the acid as the battery used in the last experiment. The weight lifted in this case was eight ounces; when the intervening wire was removed and the trough attached directly to the ends of the wire surrounding the horseshoe, it lifted only seven ounces. . . .

"It is possible that the different states of the trough with respect to dryness may have exerted some influence on this remarkable result; but that the effect of a current from a trough, if not increased, is but slightly diminished in passing through a long wire is certain. . . .

"But be this as it may, the fact that the magnetic action of a current from a trough is, at least, not sensibly diminished by passing through a long wire is directly applicable to Mr. Barlow's project of forming an electromagnetic telegraph; and it is also of material consequence in the construction of the galvanic coil. From these experiments it is evident that in forming the coil we may either use one very long wire or several shorter ones, as the circumstances may require; in the first case, our galvanic combinations must consist of a number of plates, so as to give 'projectile force;' in the second it must be formed of a single pair.

"In order to test on a large scale the truth of these preliminary results, a bar of soft iron, two inches square and 20 inches long, was bent into the form of a horseshoe $9\frac{1}{4}$ inches high. The sharp edges of the bar were first a little rounded by the hammer—it weighed 21 pounds; a piece of iron from the same bar, weighing seven pounds, was filed perfectly flat on one surface, for an armature or lifter; the extremities of the legs of the horseshoe were also truly ground to the surface of the armature; around this horseshoe 540 feet of copper bell wire were wound in nine coils of 60 feet each; these coils were not continued around the whole length of the bar, but each strand of wire, according to the principle before mentioned, occupied about two inches, and was coiled several times backward and forward over itself; the several ends of the wires were left projecting and all numbered, so that the first and last end of each strand might be readily distinguished. In this manner we formed an experimental magnet on a large scale, with which several combinations of wire could be made by merely uniting the different projecting ends. Thus if the second end of the first wire be soldered to the first end of the second wire, and so on through all the series, the whole will form a continuous coil of one long wire.

"By soldering different ends the whole may be formed in a double coil of half the length, or into a triple coil of one-third the length, etc. The horseshoe was suspended in a strong rectangular wooden frame, 3 feet 9 inches high and 20 inches wide; an iron bar was fixed below the magnet, so as to act as a lever of the second order; the different weights supported were estimated by a sliding weight in the same manner as with a common steel-yard (see sketch). In the experiments immediately following (all weights being avoirdupois) a small single battery was used, consisting of two concentric copper cylinders with zinc between them; the whole amount of zinc surface exposed to the acid from both sides of the zinc was two-fifths of a square foot; the battery required only half a pint of dilute acid for its submersion.

"*Experiment* 8.—Each wire of the horseshoe was soldered to the battery in succession, one at a time; the magnetism developed by each was just sufficient to support the weight of the armature, weighing seven pounds.

"*Experiment* 9.—Two wires, one on each side of the arch of the horseshoe, were attached; the weight lifted was 145 pounds.

"*Experiment* 10.—With two wires, one from each extremity of the legs, the weight lifted was 200 pound

"*Experiment* 11.—With three wires, one from each extremity of the legs and one from the middle of the arch, the weight supported was 300 pounds.

"*Experiment* 12.—With four wires, two from each extremity, the weight lifted was 500 pounds and the armature; when the acid was removed from the zinc, the magnet continued to support for a few minutes 130 pounds.

"*Experiment* 13.—With six wires the weight supported was 570 pounds; in all these experiments the wires were soldered to the galvanic element; the connection in no case was formed with mercury.

"*Experiment* 14.—When all the wires (nine in number) were attached, *the maximum weight lifted was 650 pounds*, and this astonishing result, it must be remembered, was produced by a battery containing only two-fifths of a square foot of zinc surface, and requiring only half a pint of dilute acid for its submersion.

"*Experiment* 15.—A small battery, formed with a plate of zinc 12 inches long and 6 inches wide, and surrounded by copper, was substituted for the galvanic elements used in the last experiment; the weight lifted in this case was 750 pounds.

"*Experiment* 16.—In order to ascertain the effect of a very small galvanic element on this large quantity of iron, a pair of plates exactly one inch square was attached to all the wires; the weight lifted was 85 pounds.

"The following experiments were made with wires of different lengths on the same horseshoe:

"*Experiment* 17.—With six wires, each 30 feet long, attached to the galvanic element, the weight lifted was 375 pounds.

"*Experiment* 18.—The same wires used in the last experiment were united so as to form three coils of 60 feet each; the weight supported was 290 pounds. This result agrees nearly with that of experiment 11, though the same individual wires were not used; from this it appears that six short wires are more powerful than three of double the length.

"*Experiment* 19.—The wires used in experiment 10, but united so as to form a single coil of 120 feet of wire, lifted 60 pounds; while in experiment 10 the weight lifted was 200 pounds. This is a confirmation of the result in the last experiment. . . .

"In these experiments a fact was observed which appears somewhat surprising: when the large battery was attached, and the armature touching both poles of the magnet, it was capable of supporting more than 700 pounds, but when only one pole was in contact it did not support more than five or six pounds, and in this case we never succeeded in making it lift the armature (weighing seven pounds). This fact may perhaps be common to all large magnets, but we have never seen the circumstance noticed of so great a difference between a single pole and both. . . .

"A series of experiments was separately instituted by Dr. Ten Eyck, in order to determine the maximum development of magnetism in a small quantity of soft iron.

"Most of the results given in this paper were witnessed by Dr. L. C. Beck, and to this gentleman we are indebted for several suggestions, and particularly that of substituting cotton well waxed for silk thread, which in these investigations became a very considerable item of expense. He also made a number of experiments with iron bonnet wires, which, being found in commerce already wound, might possibly be substituted in place of copper. The result was that with very short wire the effect was nearly the same as with copper, but in coils of long wire with a small galvanic element it was not found to answer. Dr. Beck also constructed a horseshoe of round iron one inch in diameter, with four coils on the plan before described. With one wire it lifted 30 pounds, with two wires 60 pounds, with three wires 85 pounds, and with four wires 112 pounds. While we were engaged in these investigations, the last number of the *Edinburgh Journal of Science* was received containing Professor Moll's paper on 'Electromagnetism.' Some of his results are in a degree similar to those here described; his object, however, was different, it being only to induce strong magnetism on soft iron with a powerful galvanic battery. The principal object in these experiments was to produce the greatest magnetic force with the smallest quantity of galvanism. The only effect Professor Moll's paper has had over these investigations has been to hasten their publication; the principle on which they were instituted was known to us nearly two years since, and at that time exhibited to the Albany Institute." [34]

In the next number of *Silliman's Journal* (April, 1831) Professor Henry gave "an account of a large electromagnet made for the laboratory of Yale College." The core of the armature weighed $59\frac{1}{2}$ pounds; it was forged under Henry's own direction, and wound by Dr. Ten Eyck. This magnet, wound with 26 strands of copper bell wire of a total length of 728 feet, and excited by two cells which exposed nearly $4\frac{3}{8}$ square feet of surface, readily supported on its armature, which weighed 23 pounds, a load of 2,063 pounds.

---

[34] "Scientific Writings of Joseph Henry," p. 49.

LECTURES ON THE ELECTROMAGNET. 35

Writing in 1867 of his earlier experiments, Henry

FIG. 5.—HENRY'S ELECTROMAGNET.[35]

---

[35] This figure, copied from the *Scientific American*, Dec. 11, 1880, represents Henry's electromagnet still preserved in Princeton College. The other apparatus at the foot, including a current-reverser, and the ribbon-coil used in the famous experiments on secondary and tertiary currents, were mostly constructed by Henry's own hands.

speaks[36] thus of his ideas respecting the use of additional coils on the magnet and the increase of battery power:

"To test these principles on a larger scale the experimental magnet was constructed, which is shown in Fig. 6. In this a number of compound helices were placed on the same bar, their ends left projecting, and so numbered that they could all be united into one long helix, or variously combined in sets of lesser length.

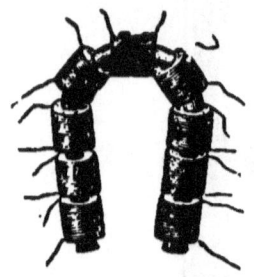

FIG. 6.—HENRY'S EXPERIMENTAL ELECTROMAGNET.

"From a series of experiments with this and other magnets, it was proved that in order to produce the greatest amount of magnetism from a battery of a single cup a number of helices is required; but when a compound battery is used then one long wire must be employed, making many turns around the iron, the length of wire, and consequently the number of turns, being commensurate with the projectile power of the battery.

"In describing the results of my experiments, the terms 'intensity' and 'quantity' magnets were introduced to avoid circumlocution, and were intended to be used merely in a technical sense. By the intensity magnet I designated a piece of soft iron, so surrounded with wire that its magnetic power could be called into operation by an intensity battery; and by a quantity magnet, a piece of iron so surrounded by a number of separate coils that its magnetism could be fully developed by a quantity battery.

"I was the first to point out this connection of the two kinds of the battery with the two forms of the magnet, in

---

[36] Statement in relation to the history of the electromagnetic telegraph, from the Smithsonian Annual Report for 1857, p. 99.

my paper, in *Silliman's Journal*, January, 1831, and clearly to state that when magnetism was to be developed by means of a compound battery one long coil must be employed, and when the maximum effect was to be produced by a single battery a number of single strands should be used. . . . Neither the electromagnet of Sturgeon nor any electromagnet ever made previous to my investigations was applicable to transmitting power to a distance. . . . The electromagnet made by Sturgeon and copied by Dana, of New York, was an imperfect quantity magnet, the feeble power of which was developed by a single battery."

Finally, Henry [37] sums up his own position as follows:

"1. Previous to my investigations the means of developing magnetism in soft iron were imperfectly understood, and the electromagnet which then existed was inapplicable to transmissions of power to a distance.

"2. I was the first to prove by actual experiment that in order to develop magnetic power at a distance a galvanic battery of 'intensity' must be employed to project the current through the long conductor, and that a magnet surrounded by many turns of one long wire must be used to receive this current.

"3. I was the first to actually magnetize a piece of iron at a distance, and to call attention to the fact of the applicability of my experiments to the telegraph.

"4. I was the first to actually sound a bell at a distance by means of the electromagnet.

"5. The principles I had developed were applied by Dr. Gale to render Morse's machine effective at a distance."

Though Henry's researches were published in 1831,

---

[37] "Scientific Writings of Joseph Henry," vol. ii., p. 435.

they were for some years almost unknown in Europe. Until April, 1837, when Henry himself visited Wheatstone at his laboratory at King's College, the latter did not know how to construct an electromagnet that could be worked through a long wire circuit. Cooke, who became the coadjutor of Wheatstone, had originally come to him to consult him,[38] in February, 1837, about his telegraph and alarum, the electromagnets of which, though they worked well on short circuits, refused to work when placed in circuit with even a single mile of wire. Wheatstone's own account [39] of the matter is extremely explicit: "Relying on my former experience, I at once told Mr. Cooke that his plan would not and could not act as a telegraph, because sufficient attractive power could not be imparted to an electromagnet interposed in a long circuit; and, to convince him of the truth of this assertion, I invited him to King's College to see the repetition of the experiments on which my conclusion was founded. He came, and after seeing a variety of voltaic magnets, which even with powerful batteries exhibited only slight adhesive traction, he expressed his disappointment."

After Henry's visit to Wheatstone, the latter altered his tone. He had been using, *faute de mieux*, relay circuits to work the electromagnets of his alarum in a short circuit with a local battery. "These short circuits," he writes, "have lost nearly all their importance

---

[38] See Mr. Latimer Clark's account of Cooke in vol. viii. of *Jour. Soc. Telegr. Engineers*, p. 374, 1880.

[39] W. F. Cooke, "The Electric Telegraph; Was it Invented by Prof. Wheatstone?" 1856–57, part ii., p. 87.

and are scarcely worth contending about since *my discovery*" (the italics are our own) "that electromagnets may be so constructed as to produce the required effects by means of the direct current, even in very long circuits." [40]

We pass on to the researches of the distinguished physicist of Manchester, whose decease we have lately had to deplore, Mr. James Prescott Joule, who, fired by the work of Sturgeon, made most valuable contributions to the subject. Most of these were published either in Sturgeon's *Annals of Electricity*, or in the *Proceedings of the Literary and Philosophical Society of Manchester*, but their most accessible form is the republished volume issued five years ago by the Physical Society of London.

In his earliest investigations he was endeavoring to work out the details of an electric motor. The following is an extract from his own account ("Reprint of Scientific Papers," p. 7):

"In the further prosecution of my inquiries, I took six pieces of round bar iron of different diameters and lengths, also a hollow cylinder, one-thirteenth of an inch thick in the metal. These were bent in the U-form, so that the shortest distance between the poles of each was half an inch; each was then wound with 10 feet of covered copper wire, one-fortieth of an inch in diameter. Their attractive powers under like currents for a straight steel magnet, 1½ inches long, suspended horizontally to the beam of a balance, were, at the distance of half an inch, as follows: (See table on page 40.)

"A steel magnet gave an attractive power of 23 grains, while its lifting power was not greater than 60 ounces.

---

[40] *Ib.*, p. 95.

	No. 1. Hollow.	No. 2. Solid.	No. 3. Solid.	No. 4. Solid.	No. 5. Solid.	No. 6. Solid.	No. 7. Solid.
Length round the bend in inches	6	5½	2⅞	5¼	2½	5¼	2¼
Diameter in inches	½	½	½	⅜	⅜	¼	¼
Attraction for steel magnet, in grains	7.5	6.3	5.1	5.0	4.1	4.8	3.6
Weight lifted, in ounces	36	52	92	36	52	20	28

"The above results will not appear surprising if we consider, first, the resistance which iron presents to the induction of magnetism, and, second, how very much the induction is exalted by the completion of the magnetic circuit.

"Nothing can be more striking than the difference between the ratios of lifting to attractive power at a distance in the different magnets. While the steel magnet attracts with a force of 23 grains and lifts 60 ounces, the electromagnet No. 3 attracts with a force of only 5.1 grains, but lifts as much as 92 ounces.

"To make a good electromagnet for lifting purposes: 1st. Its iron, if of considerable bulk, should be compound, of good quality, and well annealed. 2d. The bulk of the iron should bear a much greater ratio to its length than is generally the case. 3d. The poles should be ground quite true, and fit flatly and accurately to the armature. 4th. The armature should be equal in thickness to the iron of the magnet.

"In studying what form of electromagnet is best for attraction from a distance, two things must be considered, viz., the length of the iron and its sectional area.

"Now I have always found it disadvantageous to increase the length beyond what is needful for the winding of the covered wire."

These results were announced in March, 1839. In May of the same year he propounded a law of the mutual attraction of two electromagnets as follows: "The attractive force of two electromagnets for one another is directly proportional to the square of the electric force to which the iron is exposed; or if E denote the electric current, W the length of wire, and M the magnetic attraction, $M = E^2 W^2$." The discrepancies which he himself observed he rightly attributed to the iron becoming saturated magnetically. In March, 1840, he ex-

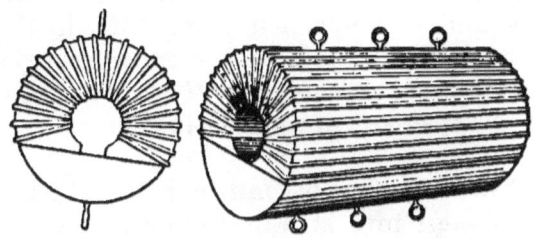

Fig. 7.—Joule's Electromagnet.

tended this same law to the lifting power of the horseshoe electromagnet.

In August, 1840, he wrote to the *Annals of Electricity* on electromagnetic forces, dealing chiefly with some special electromagnets for traction. One of these possessed the form shown in Fig. 7. Both the magnet and the iron keeper were furnished with eye-holes for the purpose of suspension and measurement of the force requisite to detach the keeper. Joule thus writes about the experiments:[41]

"I proceed now to describe my electromagnets, which I constructed of very different sizes in order to develop any

---

[41] "Scientific Papers," vol. i., p. 30.

curious circumstance which might present itself. A piece of cylindrical wrought iron, eight inches long, had a hole one inch in diameter bored the whole length of its axis, one side was planed until the hole was exposed sufficiently to separate the thus formed poles one-third of an inch. Another piece of iron, also eight inches long, was then planed, and, being secured with its face in contact with the other planed surface, the whole was turned into a cylinder eight inches long, $3\frac{3}{4}$ inches in exterior, and one inch interior diameter. The larger piece was then covered with calico and wound with four copper wires covered with silk, each 23 feet long and one-eleventh of an inch in diameter — a quantity just sufficient to hide the exterior surface, and to fill the interior opened hole. . . . The above is designated No. 1; and the rest are numbered in the order of their description.

"I made No. 2 of a bar of half-inch round iron 2.7 inches long. It was bent into an almost semicircular shape and then covered with seven feet of insulated copper wire $\frac{1}{10}$ inch thick. The poles are half an inch asunder, and the wire completely fills the space between them.

"A third electromagnet was made of a piece of iron 0.7 inch long, 0.37 inch broad, and 0.15 inch thick. Its edges were reduced to such an extent that the transverse section was elliptical. It was bent into a semicircular shape, and wound with 19 inches of silked copper wire $\frac{1}{40}$ inch in diameter.

"To procure a still more extensive variety, I constructed what might, from its extreme minuteness, be termed an *elementary electromagnet*. It is the smallest, I believe, ever made, consisting of a bit of iron wire $\frac{1}{4}$ inch long and $\frac{1}{28}$ inch in diameter. It was bent into the shape of a semicircle, and was wound with three turns of *uninsulated* copper wire $\frac{1}{40}$ inch in thickness."

With these magnets experiments were made with various strengths of currents, the tractive forces being measured by an arrangement of levers. The results, briefly, are as follows: Electromagnet No. 1, the iron of which weighed 15 pounds, required a weight of 2,090 pounds to detach the keeper. No. 2, the iron of which weighed 1,057 grains, required 49 pounds to detach its armature. No. 3, the iron of which weighed 65.3 grains, supported a load of 12 pounds, or 1,286 times its own weight. No. 4, the weight of which was only half a grain, carried in one instance 1,417 grains, or 2,834 times its own weight.

"It required much patience to work with an arrangement so minute as this last; and it is probable that I might ultimately have obtained a larger figure than the above, which, however, exhibits a power proportioned to its weight far greater than any on record, and is eleven times that of the celebrated steel magnet which belonged to Sir Isaac Newton.

"It is well known that a steel magnet ought to have a much greater length than breadth or thickness; and Mr. Scoresby has found that when a large number of straight steel magnets are bundled together, the power of each when separated and examined is greatly deteriorated. All this is easily understood, and finds its cause in the attempt of each part of the system to induce upon the other part a contrary magnetism to its own. Still there is no reason why the principle should in all cases be extended from the steel to the electromagnet, since in the latter case a great and commanding inductive power is brought into play to sustain what the former has to support by its own unassisted retentive property. All the preceding experiments support this position; and the following table gives proof of the obvious and necessary general consequence: the maximum power of the electromagnet is directly proportional to its least transverse sectional area. The second column of the table contains the least sectional area in square inches of the entire magnetic circuit. The maximum power in pounds avoirdupois is recorded in the third; and this, reduced to an inch square of sectional area, is given in the fourth column under the title of specific power.

TABLE I.

DESCRIPTION.	Least sectional area.	Maximum power.	Specific power.
My own electromagnets No. 1	10	2,090	209
No. 2	0.196	49	250
No. 3	0.0436	12	275
No. 4	0.0012	0.202	162
Mr. J. C. Nesbit's.. Length round the curve, 3 feet; diameter of iron core, 2¾ inches; sectional area, 5.7 inches; do. of armature, 4.5 inches; weight of iron, about 50 pounds	4.5	1,428	317
Prof. Henry's. Length round the curve, 20 inches; section, 2 inches square; sharp edges rounded off; weight, 21 pounds	3.94	750	190
Mr. Sturgeon's original. Length round the curve, about 1 foot; diameter of the round bar, ½ inch	0.196	50	255

"The above examples are, I think, sufficient to prove the rule I have advanced. No. 1 was probably not fully saturated; otherwise I have no doubt that its power per square inch would have approached 300. Also the specific power of No. 4 is small, because of the difficulty of making a good experiment with it."

These experiments were followed by some to ascertain the effect of the length of the iron of the magnet, which he considered, at least in those cases where the degree of magnetization is considerably below the point of saturation, to offer a decidedly proportional resistance to magnetization; a view the justice of which is now, after 50 years, amply confirmed.

In November of the same year further experiments [42] in the same direction were published. A tube of iron, spirally made and welded, was prepared, planed down as in the preceding case, and fitted to a similarly prepared armature. The hollow cylinder thus formed, shown in Fig. 8, was two feet in length. Its external diameter was 1.42 inches, its internal being 0.5 inch. The least sectional area was $10\frac{1}{4}$ square inches. The exciting coil consisted of a single copper rod, covered with tape, bent into a sort of S-shape. This was later replaced by a coil of 21 copper wires, each $\frac{1}{25}$ inch in

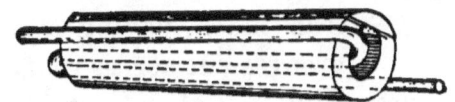

Fig. 8.—Joule's Cylindrical Electromagnet.

diameter and 23 feet long, bound together by cotton tape. This magnet, excited by a battery of 16 of Sturgeon's cast-iron cells, each one foot square and one and a half inches in interior width, arranged in a series of four, gave a lifting power of 2,775 pounds.

Joule's work was well worthy of the master from whom he had learned his first lesson in electromagnetism. He showed his devotion not only by writing descriptions of them for Sturgeon's *Annals*, but by exhibiting two of his electromagnets at the Victoria Gallery of Practical Science, of which Sturgeon was director. Others, stimulated into activity by Joule's example, proposed new forms, among them being two Manchester

---

[42] "Scientific Papers," p. 40, and *Annals of Electricity*, vol. v., p. 170.

gentlemen, Mr. Radford and Mr. Richard Roberts, the latter being a well-known engineer and inventor. Mr. Radford's electromagnet consisted of a flat iron disc with deep spiral grooves cut in its face, in which were laid the insulated copper wires. The armature consisted of a plain iron disc of similar size. This form is described in Vol. IV. of Sturgeon's *Annals*.

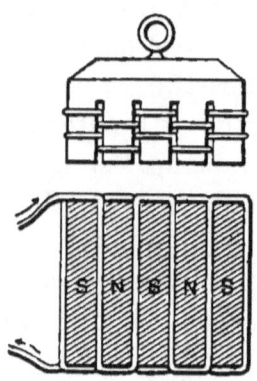

FIG. 9.—ROBERTS' ELECTROMAGNET.

Mr. Roberts' form of electromagnet consisted of a rectangular iron block, having straight parallel grooves cut across its face, as in Fig. 9. This was described in Vol. VI. of Sturgeon's *Annals*, page 166. Its face was $6\frac{5}{8}$ inches square and its thickness $2\frac{7}{16}$ inches. It weighed, with the conducting wire, 35 pounds; and the armature, of the same size and $1\frac{1}{2}$ inches thick, weighed 23 pounds. The load sustained by this magnet was no less than 2,950 pounds. Roberts inferred that a magnet if made of equal thickness, but five feet square, would sustain 100 tons' weight. Some of Roberts' apparatus is still preserved in the Museum of Peel Park, Manchester.

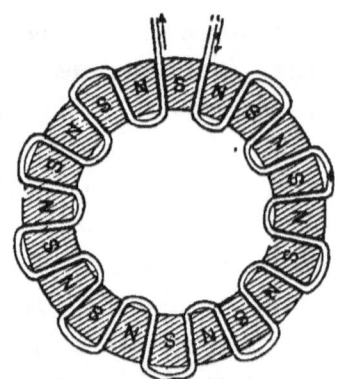

FIG. 10.—JOULE'S ZIGZAG ELECTROMAGNET.

On page 431 of the same volume of the *Annals* Joule

described yet another form of electromagnet, the form of which resembled in general Fig. 10, but which, in actual fact, was built up of 24 separate flat pieces of iron bolted to a circular brass ring. The armature was a similar structure, but not wound with wire. The iron of the magnet weighed seven pounds and that of the armature 4.55 pounds. The weight lifted was 2,710 pounds when excited by 16 of Sturgeon's cast-iron cells.

In a subsequent paper on the calorific effects of magneto-electricity,[43] published in 1843, Joule described another form of electromagnet of horseshoe shape, made from a piece of boiler-plate. This was not intended to give great lifting power, and was used as the field magnet of a motor. In 1852 another powerful electromagnet of horseshoe form, somewhat similar to the preceding, was constructed by Joule for experiment. He came to the conclusion [44] that, owing to magnetic saturation setting in, it was improbable that any force of electric current could give a magnetic attraction greater than 200 pounds per square inch. "That is, the greatest weight which could be lifted by an electromagnet formed of a bar of iron one inch square, bent into a semicircular shape, would not exceed 400 pounds."

With the researches of Joule may be said to end the first stage of development. The notion of the magnetic circuit which had thus guided Joule's work did not commend itself at that time to the professors of physical theories; and the practical men, the telegraph en-

---

[43] "Scientific Papers," vol. i., p. 123; and *Phil. Mag.*, ser. iii., vol. xxiii., p. 263, 1843.
[44] "Scientific Papers," vol. i., p. 362; and *Phil. Mag.*, ser. iv., vol. iii., p. 32.

gineers, were for the most part content to work by purely empirical methods. Between the practical man and the theoretical man there was, at least on this topic, a great gulf fixed. The theoretical man, arguing as though magnetism consisted in a surface distribution of polarity, and as though the laws of electromagnets were like those of steel magnets, laid down rules not applicable to the cases which occur in practice, and which hindered rather than helped progress. The practical man, finding no help from theory, threw it on one side as misleading and useless. It is true that a few workers made careful observations and formulated into rules the results of their investigations. Among these, the principal were Ritchie, Robinson, Müller, Dub, Von Kolke, and Du Moncel; but their work was little known beyond the pages of the scientific journals wherein their results were described. Some of these results will be examined in my later lectures, but they cannot be discussed in this historical *résumé*, which is accordingly closed.

GENERALITIES CONCERNING ELECTROMAGNETS.

Materials.—In any complete treatise on the electromagnet it would be needful to enumerate and to discuss in detail the several constructive features of the apparatus. Three classes of material enter into its construction: first, the iron which constitutes the material of the magnetic circuit, including the armature as well as the cores on which the coils are wound, and the yoke that connects them; secondly, the copper which is employed as the material to conduct the electric currents, and which is usually in the form of wire; thirdly, the insulating material employed to prevent the copper coils from coming into contact with one another, or with the iron core. There is a further subject for discussion in the bobbins, formers, or frames upon which the coils are in so many cases wound, and which may in some cases be made in metal, but often are not. The engineering of the electromagnet might well furnish matter for a special chapter.

TYPICAL FORMS.

It is difficult to devise a satisfactory or exhaustive classification of the varied forms which the electromagnet has assumed, but it is at least possible to enumerate some of the typical forms.

1. *Bar Electromagnet.*—This consists of a single straight core (whether solid, tubular, or laminated), surrounded by a coil. Fig. 3 depicted Sturgeon's earliest example.

2. *Horseshoe Electromagnet.*—There are two sub-types included in this name. The original electromagnet of Sturgeon (Fig. 1) really resembled a horseshoe in form, being constructed of a single piece of round wrought iron, about half an inch in diameter and nearly a foot long, bent into an arch. In recent years the other sub-type has prevailed, consisting, as shown in Fig. 11, of two separate iron cores, usually cut from a circular rod, fixed into a third piece of wrought iron, the yoke. Occasionally this form is modified by the use of one coil only, the second core being left uncovered. This form has received in France the name of *aimant boiteux*. Its

merits will be considered later. Sometimes a single coil is wound upon the yoke, the two limbs being uncovered.

FIG. 11.—TYPICAL TWO-POLE ELECTROMAGNET.

3. *Iron-clad Electromagnet.*—This form, which has many times been re-invented, differs from the simple bar magnet in having an iron shell or casing external to the coils and attached to the core at one end. Such a magnet presents, as depicted in Fig. 12, a central pole at one end surrounded by an outer annular pole of the opposite polarity. The appropriate armature for electromagnets of this type is a circular disc or lid of iron.

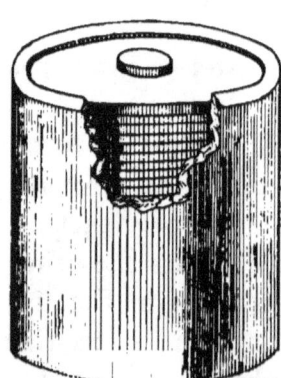

FIG. 12.—IRON-CLAD ELECTROMAGNET.

4. *Coil-and-Plunger.* — A detached iron core is attracted into a hollow coil, or solenoid, of copper wire, when a cur-

rent of electricity flows round the latter. This is a special form, and will receive extended consideration.

5. *Special Forms.*—Besides the leading forms enumerated above, there are a number of special types, multipolar, spiral, and others designed for particular purposes. There is also a group of forms intermediate between the ordinary electromagnet and the coil-and-plunger form.

POLARITY.

It is a familiar fact that the polarity of an electromagnet depends upon the sense in which the current is flowing around it. Various rules for remembering

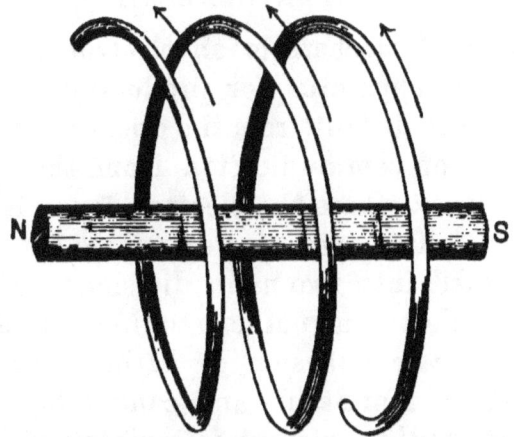

FIG. 13.—DIAGRAM ILLUSTRATING RELATION OF MAGNETIZING CIRCUIT AND RESULTING MAGNETIC FORCE.

the relation of the electric flow and the magnetic force have been given. One of them that is useful is that when one is looking at the north pole of an electromagnet, the current will be flowing around that pole in the sense opposite to that in which the hands of a clock are seen to revolve. Another useful rule, suggested by Maxwell, is illustrated by Fig. 13, namely, that the sense of the circulation of the current (whether right or left handed) and the positive direction of the resulting magnetic force are related together in the same way as the rotation and the travel of a right-handed screw are associated together. Right-handed rotation of the screw is associated with forward travel. Right-handed circulation of a current is associated with a magnetic force tending to produce north polarity at the forward end of the core.

USES IN GENERAL.

As a piece of mechanism an electromagnet may be regarded as an apparatus for producing a mechanical action at a place distant from the operator who controls it, the means of communication from the operator to the distant point where the electromagnet is being the electric wire. The uses of electromagnets may, however, be divided into two main divisions. For certain purposes an electromagnet is required merely for obtaining temporary adhesion or lifting power. It attaches itself to an armature and cannot be detached so long as the exciting current is maintained, except by the application of a superior opposing pull. The force which an electromagnet thus exerts upon an armature of iron, with which it is in direct contact, is always considerably greater than the force with which it can act on an armature at some distance away, and the two cases must be carefully distinguished. *Traction* of an armature in contact and *attraction* of an armature at a distance are two different functions. So different, indeed, that it is no exaggeration to say that an electromagnet designed for the one purpose is unfitted for the other. The question of designing electromagnets for either of these purposes will occupy a large part of these lectures. The action which an electromagnet exercises on an armature in its neighborhood may be of several kinds. If the armature is of soft iron, placed nearly parallel to the polar surfaces, the action is one simply of attraction, producing a motion of pure translation, irrespective of the polarity of the magnet. If the armature lies oblique to the lines of the poles there will be a tendency to turn it round, as well as to attract it; but, again, if the armature is of soft iron the action will be independent of the polarity of the magnet, that is to say, independent of the direction of the exciting current. If, however, the armature be itself a magnet of steel permanently magnetized, then the direction in which it tends to turn, and the amount, or even the sign of the force with which it is attracted, will depend on the polarity of the electromagnet; that is to say, will depend on the direction in which the exciting current circulates. Hence there arises a difference between the operation of a *non-polarized* and that of a *polarized* apparatus, the latter term being applied to those forms in which there is employed a portion—say an armature— to which an initial fixed magnetization has been imparted. Non-polarized apparatus is in all cases independent of the direction of the current. Another class of uses served by electromagnets is the production of rapid vibrations. These are employed in the mechanism of electric trembling bells, in the automatic breaks of induction coils, in electrically driven tuning-forks such as are employed for chronographic purposes, and in the instruments used in harmonic telegraphy. Special constructions of electromagnets are appropriate to special purposes such as these. The adaptation of electromagnets for the special end of responding to rapidly alternating currents is a closely kindred matter. Lastly, there are certain applications of the electromagnet, notably in the construction of some forms of arc lamp, for which it is specially sought to obtain an equal, or approximately equal, pull over a definite range of motion. This use necessitates special designs.

THE PROPERTIES OF IRON.

A knowledge of the magnetic properties of iron of different kinds is absolutely fundamental to the theory and design of electromagnets. No excuse is therefore necessary for treating this matter with some fullness. In all modern treatises on magnetism the usual terms are defined and explained. Magnetism, which was formerly treated of as though it were something distributed over the end surfaces of magnets, is now known to be a phenomenon of internal structure; and the appropriate mode of considering it is to treat the magnetic materials, iron and the like, as being capable of acting as good conductors of the magnetic lines; in other words, as possessing magnetic *permeability*. The precise notion now attached to this word is that of a numerical coefficient. Suppose a magnetic force—due, let us say, to the circulation of an electric current in a surrounding coil—were to act on a space occupied by air: there would result a certain number of magnetic lines in that space. In fact, the intensity of the magnetic force, symbolized by the letter **H**, is often expressed by saying that it would produce **H** magnetic lines per square centimetre in air. Now, owing to the superior magnetic power of iron, if the space subjected to this magnetic force were filled with iron instead of air, there would be produced a larger number of magnetic lines per square centimetre. This larger number in the iron expresses the degree of magnetization in the iron; it is symbolized [45] by the letter **B**. The ratio of **B** and **H** expresses the permeability of the material. The usual symbol for permeability is the Greek letter μ. So we may say that **B** is equal to μ times **H**. For example, a certain specimen of iron when subjected to a magnetic force capable of creating, in air, 50 magnetic lines to the square centimetre, was found to be permeated by no fewer than 16,062 magnetic lines per square

---

[45] The following are the various ways of expressing the three quantities under consideration:

B—The internal magnetization.
　The magnetic induction.
　The induction.
　The intensity of the induction.
　The permeation.
　The number of lines per square centimetre in the material.

H—The magnetizing force at a point.
　The magnetic force at a point.
　The intensity of the magnetic force.
　The number of lines per square centimetre that there would be in air.

μ—The magnetic permeability.
　The permeability.
　The specific conductivity for magnetic lines.
　The magnetic multiplying power of the material.

centimetre. Dividing the latter figure by the former gives as the value of the permeability at this stage of the magnetization 321, or the permeability of the iron is 321 times that of air. The permeability of such non-magnetic materials as silk, cotton, and other insulators, also of brass, copper, and all the non-magnetic metals, is taken as 1, being practically the same as that of the air.

This mode of expressing the fact is, however, complicated by the fact of the tendency in all kinds of iron to magnetic saturation. In all kinds of iron the magnetizability of the material becomes diminished as the actual magnetization is pushed further. In other words, when a piece of iron has been magnetized up to a certain degree it becomes, from that degree onward, less permeable to further magnetization, and though actual saturation is never reached, there is a practical limit beyond which the magnetization cannot well be pushed. Joule was one of the first to establish this tendency toward magnetic saturation. Modern researches have shown numerically how the permeability diminishes as the magnetization is pushed to higher stages. The practical limit of the magnetization, **B**, in good wrought iron is about 20,000 magnetic lines to the square centimetre, or about 125,000 lines to the square inch; and in cast iron the practical saturation limit is nearly 12,000 lines per square centimetre, or about 70,000 lines per square inch. In designing electromagnets, before calculations can be made as to the size of a piece of iron required for the core of a magnet for any particular purpose, it is necessary to know the magnetic properties of that piece of iron; for it is obvious that if the iron be of in-

LECTURES ON THE ELECTROMAGNET. 57

ferior magnetic permeability, a larger piece of it will be required in order to produce the same magnetic effect as might be produced with a smaller piece of higher permeability. Or, again, the piece having inferior permeability will require to have more copper wire wound on it; for in order to bring up its magnetization to the required point, it must be subjected to higher magnetiz-

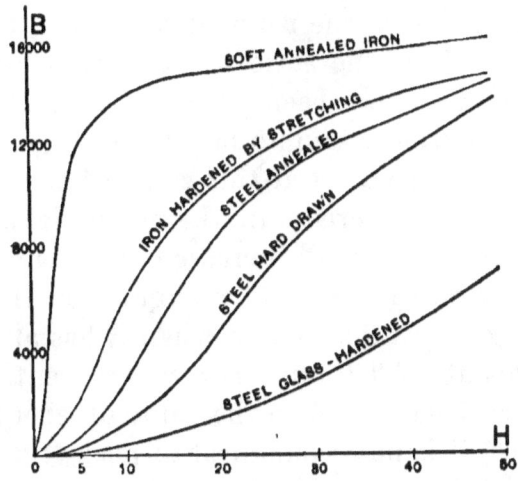

FIG. 14.—CURVES OF MAGNETIZATION OF DIFFERENT MAGNETIC MATERIALS.

ing forces than would be necessary if a piece of higher permeability had been selected.

A convenient mode of studying the magnetic facts respecting any particular brand of iron is to plot on a diagram the curve of magnetization—*i. e.*, the curve in which the values, plotted horizontally, represent the magnetic force **H**, and the values plotted vertically those that correspond to the respective magnetization **B**. In Fig. 14, which is modified from the researches of Prof.

Ewing, are given five curves relating to soft iron, hardened iron, annealed steel, hard drawn steel, and glass-hard steel. It will be noticed that all these curves have the same general form. For small values of **H** the values of **B** are small, and as **H** is increased **B** increases also. Further, the curve rises very suddenly, at least with all the softer sorts of iron, and then bends over and becomes nearly horizontal. When the magnetization is in the stage below the bend of the curve, the iron is said to be far from the state of saturation. But when the magnetization has been pushed beyond the bend of the curve, the iron is said to be in the stage approaching saturation; because at this stage of magnetization it requires a large increase in the magnetizing force to produce even a very small increase in the magnetization. It will be noted that for soft wrought iron the stage of approaching saturation sets in when **B** has attained the value of about 16,000 lines per square centimetre, or when **H** has been raised to the value of about 50. As we shall see, it is not economical to push **B** beyond this limit; or, in other words, it does not pay to use stronger magnetic forces than those of about **H** $= 50$.

METHODS OF MEASURING PERMEABILITY.

There are four sorts of experimental methods of measuring permeability.

1. *Magnetometric Methods.*—These are due to Müller, and consist in surrounding a bar of the iron in question by a magnetizing coil and observing the deflection its magnetization produces in a magnetometer.

2. *Balance Methods.*—These methods are a variety of

the preceding, a compensating magnet being employed to balance the effect produced by the magnetized iron on the magnetometric needle. Von Feilitzsch used this method, and it has received a more definite application in the magnetic balance of Prof. Hughes. The actual balance is exhibited to-night upon the table, and I have beside me a large number of observations made by students of the Finsbury Technical College by its means upon sundry samples of iron and steel. None of these methods are, however, to be compared with those that follow.

3. *Inductive Methods.*—There are several varieties of these, but all depend on the generation of a transient induction current in an exploring coil which surrounds the specimen of iron, the integral current being proportional to the number of magnetic lines introduced into, or withdrawn from, the circuit of the exploring coil. Three varieties may be mentioned.

(*A*) *Ring Method.*—In this method, due to Kirchhoff, the iron under examination is made up into a ring, which is wound with a primary or exciting coil and with a secondary or exploring coil. Determinations on this plan have been made by Stowletow, Rowland, Bosanquet, and Ewing; also by Hopkinson. Rowland's arrangement of the experiment is shown in Fig. 15 in which B is the exciting battery; S, the switch for turning on or reversing the current; R, an adjustable resistance; A, an ampèremeter; and $B\,G$ the ballistic galvanometer, the first swing of which measures the integral induced current. $R\,C$ is an earth inductor or reversing coil wherewith to calibrate the readings of the galvanometer; and above is an arrangement of a coil and a magnet to assist in bringing the swinging needle to rest between the observations. The exciting coil and the exploring coil are both wound upon the ring: the former is distinguished by being drawn with a thicker line. The usual mode of procedure is to begin with a feeble exciting current, which is suddenly reversed, and then reversed back. The current is then increased, reversed

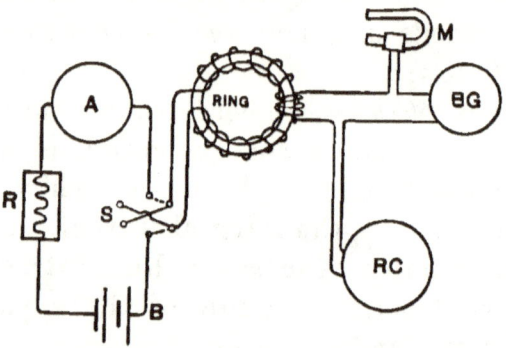

Fig. 15.—Ring Method of Measuring Permeability (Rowland's Arrangement).

and re-reversed; and so on, until the strongest available points are reached. The values of the magnetizing force **H** are calculated from the observed value of the current by the following rule. If the strength of the current, as measured by the ampèremeter, be i, the number of spires of the exciting coil S and the length, in centimetres, of the coil (*i. e.*, the mean circumference of the ring) be l, then **H** is given by the formula:

$$\mathbf{H} = \frac{4\pi}{10} \times \frac{Si}{l} = 1.2566 \times \frac{Si}{l}$$

Bosanquet, applying this method to a number of iron rings, obtained some important results.

In Fig. 16 are plotted out the values of **H** and **B** for seven rings. One of these, marked J, was of cast steel, and was examined both when soft and afterward when hardened. Another, marked I, was of the best Lowmoor iron. Five were of Crown iron, of different sizes. They were marked for distinction with the letters G, E, F, H, K. In the accompanying table are set down the values of **B** at different stages of the magnetization.

TABLE OF VALUES OF **B** IN FIVE CROWN IRON RINGS.

Name.	G.	E.	F.	H.	K.
Mean Diameter.	21.5 cm.	10.035 cm.	22.1 cm.	10.735 cm.	22.725 cm.
Bar thickness.	2.535	1.298	1.292	0.7137	0.7554
Magnetizing Force.					
0.2	126	73	62	82	85
0.5	377	270	224	208	214
1	1,449	1,293	840	675	885
2	4,564	3,952	3,533	2,777	2,417
5	9,900	9,147	8,293	8,479	8,884
10	13,023	13,357	12,540	11,376	11,388
20	14,911	14,653	14,710	14,006	13,273
50	16,217	15,704	16,062	15,174	13,890
100	17,148	16,677	17,900	16,134	14,837

I have the means here of illustrating the induction method of measuring permeability. Here is an iron ring, having a cross-section of almost exactly one square centimetre. It is wound with an exciting coil supplied with current by two accumulator cells; over it is also wound an exploring coil of 100 turns connected in circuit (as in Rowland's arrangement) with a ballistic galvanometer which reflects a spot of light upon yonder screen. In the circuit of the galvanometer is also included a reversing earth coil. As a matter of fact this earth coil is of such a size, and wound with so many convolutions of wire, that when it is turned over the amount of cutting of magnetic lines is equal to 840,000, or is the same as if 840,000 magnetic lines had been cut once. By adjusting the resistance of the galvanometer circuit, it is arranged that the first swing due to the induced current when I suddenly turn over the earth

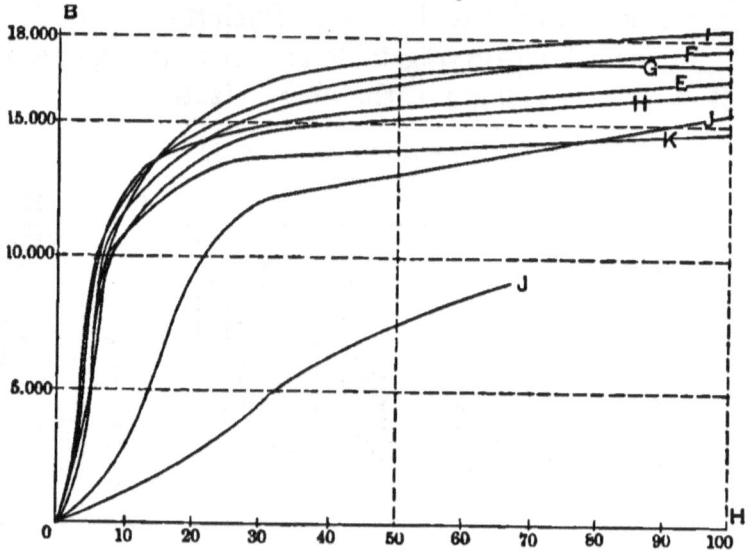

FIG. 16.—BOSANQUET'S DATA OF MAGNETIC PROPERTIES OF IRON AND STEEL RINGS.

coil is 8.4 scale divisions. Then, seeing that our exploring coil has 100 turns, it follows that when in our subsequent experiment with the ring we get an induced current from it, each division of the scale over which the spot swings will mean 1,000 lines in the iron. I turn on my exciting current. See: it swings about 11 divisions. On breaking the circuit it swings nearly 11 divisions the other way. That means that the magnetizing force carries the magnetization of the iron up to 11,000 lines; or, as the cross-section is about one square centimetre, $B = 11,000$. Now, how much is H? The exciting coil has 180 windings, and the exciting current through the ampèremeter is just one ampère. The total excitation is just 180 "ampère turns." We must, according to our rule given above, multiply this by 1.2566 and divide by the mean circumferential length of the coil, which is about 32 centimetres. This makes $H = 7$. So if $B = 11,000$ and $H = 7$, the permeability (which is the ratio of them) is about 1,570. It is a rough and hasty experiment, but it illustrates the method.

Bosanquet's experiments settled the debated question whether the outer layers of an iron core shield the inner layers from the influence of magnetizing forces. Were this the case, the rings made from thin bar iron should exhibit higher values of B than do the thicker rings. This is not so; for the thickest ring, G, shows throughout the highest magnetizations.

(*B*) *Bar Method.*—This method consists in employing a long bar of iron instead of a ring. It is covered from end to end with the exciting coil, but the exploring coil consists of but a few turns of wire situated just over the middle part of the bar. Rowland, Bosanquet, and Ewing have all employed this variety of method; and Ewing specially used bars, the length of which was more than 100 times their diameter, in order to get rid of errors arising from end effects.

(*C*) *Divided Bar Method.*—This method, due to Dr. Hopkinson,[46] is illustrated by Fig. 17.

---

[46] *Phil. Trans.*, 1885, p. 504.

The apparatus consists of a block of annealed wrought iron about 18 inches long, 6½ wide, and 2 deep, out of the middle of which is cut a rectangular space to receive the magnetizing coils.

The test samples of iron consist of two rods, each 12.65 millimetres in diameter, turned carefully true, which slide in through holes bored in the ends of the iron blocks. These two rods meet in the middle, their ends

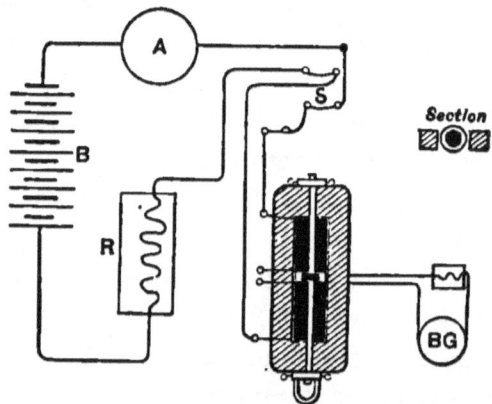

Fig. 17.—Hopkinson's Divided Bar Method of Measuring Magnetic Permeability.

being faced true so as to make a good contact. One of them is secured firmly, and the other has a handle fixed to it, by means of which it can be withdrawn. The two large magnetizing coils do not meet, a space being left between them. Into this space is introduced the little exploring coil, wound upon an ivory bobbin, through the eye of which passes the end of the movable rod. The exploring coil is connected to the ballistic galvanometer, $B\ G$, and is attached to an india-rubber spring (not shown in the figure), which, when the rod is suddenly pulled back, causes it to leap entirely out of the magnetic field. The exploring coil had 350 turns of fine wire; the two magnetizing coils had 2,008 effective turns. The magnetizing current, generated by a battery, B, of eight Grove cells, was regulated by a variable liquid resistance, R, and by a shunt resistance. A reversing switch and an ampèremeter, A, were included in the magnetizing circuit. By means of this apparatus the sample rods to be experimented upon could be submitted to any magnetizing forces, small or large, and the actual magnetic condition could be examined at any time by breaking the circuit and simultaneously withdrawing the movable rod. This apparatus, therefore, permitted the observation separately of a series of increasing (or decreasing) magnetizations without any intermediate reversals of the entire current. Thirty-five samples of various irons of known chemical composition were examined by Hopkinson, the two most important for present purposes being an annealed wrought iron and a gray cast iron, such as are used by Messrs. Mather and Platt in the construction of dynamo machines. Hopkinson embodied his results in curves, from which it is possible to construct, for purposes of reference, numerical tables of sufficient accuracy to serve for future calculations. The curves of these two samples of iron are reproduced in Fig. 18, but with one simple modification. British engineers, who unfortunately are condemned by local circumstances to use inch measures instead of the international metric system, prefer to have the magnetic facts also stated in terms of square inch units instead of square centimetre units. This change has been made in Fig. 18, and the symbols B_u and H_u are chosen to indicate the numbers of magnetic lines to the square inch in iron and in air respectively. The permeability or multiplying power of the iron is

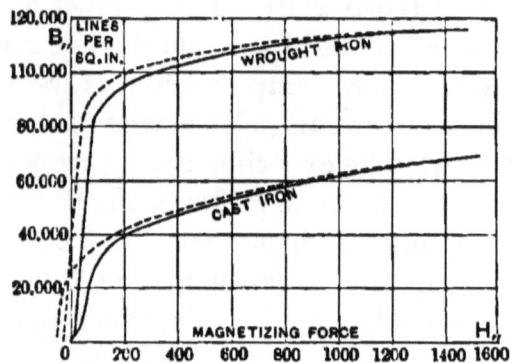

FIG. 18.—CURVES OF MAGNETIZATION OF IRON.

the same, of course, in either measure. In Table II. are given the corresponding data in square inch measure, and in Table III. the data in square centimetre measure for the same specimens of iron.

TABLE II. (Square Inch Units.)

Annealed Wrought Iron.			Gray Cast Iron.		
B_u	μ	H_u	B_u	μ	H_u
30,000	4,650	6.5	25,000	763	32.7
40,000	3,877	10.3	30,000	756	39.7
50,000	3,031	16.5	40,000	258	155
60,000	2,159	27.8	50,000	114	439
70,000	1,921	36.4	60,000	74	807
80,000	1,409	56.8	70,000	40	1,480
90,000	907	99.2			
100,000	408	245			
110,000	166	664			
120,000	76	1,581			
130,000	35	3,714			
140,000	27	5,185			

TABLE III. (Square Centimetre Units.)

Annealed Wrought Iron.			Gray Cast Iron.		
B	μ	H	B	μ	H
5,000	3,000	1.66	4,000	800	5
9,000	2,250	4	5,000	500	10
10,000	2,000	5	6,000	279	21.5
11,000	1,692	6.5	7,000	133	42
12,000	1,412	8.5	8,000	100	80
13,000	1,083	12	9,000	71	127
14,000	823	17	10,000	53	188
15,000	526	28.5	11,000	37	292
16,000	320	50			
17,000	161	105			
18,000	90	200			
19,000	54	350			
20,000	30	666			

It will be noted that Hopkinson's curves are double, there being one curve for the ascending magnetizations and a separate one, a little above the former, for descending magnetizations. This is a point of a little importance in designing electromagnets. Iron, and particularly hard sorts of iron, and steel, after having been subjected to a high degree of magnetizing force and subsequently to a lesser magnetizing force, are found to retain a higher degree of magnetization than if the lower magnetizing force had been simply applied. For example, reference to Fig. 18 shows that the wrought iron, where subjected to a magnetizing force gradually rising from zero to $H_u = 200$, exhibits a magnetization of $B_u = 95,000$; but after H_u has been carried up to over 1,000 and then reduced again to 200, B_u does not come down again to 95,000, but only to 98,000. Any sample of iron which showed great retentive qualities, or in which the descending curve differs widely from the ascending curve, would be unsuitable for constructing electromagnets, for it is important that there should be as little residual magnetism as possible in the cores. It will be noted that the curves for cast iron show more of this residual effect than do those for wrought iron. The numerical data in Tables II. and III. are means between the ascending and descending values.

As an example of the use of the Tables we may take the following: How strong must the magnetizing force be in order to produce in wrought iron a magnetization of 110,000 lines to the square inch? Reference to Table II. or to Fig. 18 shows that a magnetizing field of 664 will be required, and that at this stage of the magnetization the permeability of the iron is only 166. As there are 6.45 square centimetres to the square inch, 110,000 lines to the square inch corresponds very nearly to 17,000 lines to the square centimetre, and $H_\prime = 664$ corresponds very nearly to $H = 100$.

TRACTION METHODS.

Another group of the methods of measuring permeability is based upon the law of magnetic traction. Of these there are several varieties.

(*D*) *Divided Ring Method.*—Mr. Shelford Bidwell has kindly lent me the apparatus with which he carried out this method. It consists of a ring of very soft charcoal iron rod 6.4 millimetres in thickness, the external diameter being eight centimetres, sawn into two half rings, and then each half carefully wound over with an exciting coil of insulated copper wire of 1,929 convolutions in total. The two halves fit neatly together; and in this position it constitutes practically a continuous ring. When an exciting current is passed round the coils both halves become magnetized and attract one another. The force required to pull them asunder is then measured. According to the law of traction, which will occupy us in the second lecture, the tractive force (over a given area of contact) is proportional to the square of the number of magnetic lines that pass from one surface to the other through the contact joint. Hence the force of traction may be used to determine **B**; and on calculating **H** as before we can determine the permeability. The following Table IV. gives a summary of Mr. Bidwell's results:

TABLE IV. (Square Centimetre Measure.) Soft Charcoal Iron.

B	μ	H
7,390	1899.1	3.9
11,550	1121.4	10.3
15,460	386.4	40
17,330	150.7	115
18,470	88.8	208
19,330	45.3	427
19,820	33.9	585

(*E*) *Divided Rod Method.*—In this method, also used by Mr. Bidwell, an iron rod hooked at both ends was divided across the middle, and placed within a vertical surrounding magnetizing coil. One hook was hung up to an overhead support; to the lower hook was hung a scale pan. Currents of gradually increasing strength were sent around the magnetizing coil from a battery of cells, and note was taken of the greatest weight which could in each case be placed in the scale pan without tearing asunder the ends of the rods.

(*F*) *Permeameter Method.*—This is a method which I have myself devised for the purpose of testing specimens of iron. It is essentially a workshop method, as distinguished from a laboratory method. It requires no ballistic galvanometer, and the iron does not need to be forged into a ring or wound with a coil. For carrying it out a simple instrument is needed, which I venture to denominate as a *permeameter.* Outwardly, it has a general resemblance to Dr. Hopkinson's apparatus, and consists, as you see (Fig. 19), of a rectangular piece of soft wrought iron, slotted out to receive a magnetizing coil, down the axis of which passes a brass tube. The block is 12 inches long, 6½ inches wide, and 3 inches in thickness. At one end the block is bored to receive the sample of iron that is to be tested. This consists simply of a thin rod about a foot long, one end of which must be carefully surfaced up. When it is placed inside the magnetizing coil and the exciting current is turned on, the rod sticks tightly at its lower end to the surface of the iron block; and the force required to detach it (or, rather, the square root of that force) is a measure of the permeation of the magnetic lines through its end face. In the first permeameter which I constructed the magnetizing coil is 13.64 centimetres in length and has

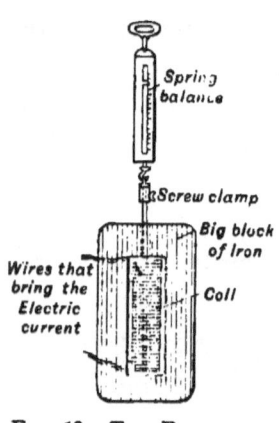

FIG. 19.—THE PERMEAMETER.

371 turns of wire. One ampère of exciting current consequently produces a magnetizing force of $H = 34$. The wire is thick enough to carry 30 ampères, so that it is easy to reach a magnetizing force of 1,000. The current I now turn on is 25 ampères. The two rods here are of "charcoal iron" and "best iron" respectively; they are of quarter-inch square stuff. Here is a spring balance graduated carefully, and provided with an automatic catch so that its index stops at the highest reading. The tractive force of the charcoal iron is about 12½ pounds, while that of the "best" iron is only 7½ pounds. B is about 19,000 in the charcoal iron, and H being 850, μ is about 22.3. The law of traction which I use in calculating B will occupy us much in the next lecture; but meantime I content myself in stating it here for use with the permeameter. The formula for calculating B when the core is thus detached by a pull of P pounds, the area of contact being A square inches, is as follows:

$$B = 1,317 \times \sqrt{P \div A} + H.$$

I may add that the instrument, in its final form, was manufactured from my designs by Messrs. Nalder Bros., the well-known makers of so many electrical instruments.

CURVES OF MAGNETIZATION AND PERMEABILITY.

In reviewing the results obtained, it will be noted that the curves of magnetization all possess the same general features, all tending toward a practical maximum, which, however, is different for different materials. Joule expressed the opinion that "no force of current could give an attraction equal to 200 pounds per square inch," the greatest he actually attained being only 175 pounds per square inch. Rowland was of opinion that the limit was about 177 pounds per square inch for an ordinary good quality of iron, even with infinitely great exciting power. This would correspond roughly to a limiting value of **B** of about 17,500 lines to the square centimetre. This value has, however, been often surpassed. Bidwell obtained 19,820, or possibly a trifle more, as in Bidwell's calculation the value of **H** has been needlessly discounted. Hopkinson gives 18,250 for wrought iron and 19,840 for mild Whitworth steel. Kapp gives 16,740 for wrought iron, 20,460 for charcoal iron in sheet, and 23,250 for charcoal iron in wire. Bosanquet found the highest value in the middle bit of a long bar to run up in one specimen to 21,428, in another to 29,388, in a third to 27,688. Ewing, working with extraordinary magnetic power, forced up the value of **B** in Lowmoor iron to 31,560 (when μ came down to 3), and subsequently to 45,350. This last figure corresponds to a traction exceeding 1,000 pounds to the square inch.

Cast iron falls far below these figures. Hopkinson, using a magnetizing force of 240, found the values of **B** to be 10,783 in gray cast iron, 12,408 in malleable cast iron, and 10,546 in mottled cast iron. Ewing, with a magnetizing force nearly 50 times as great, forced up the value of **B** in cast iron to 31,760. Mitis metal, which is a sort of cast wrought iron, being a wrought iron rendered fluid by addition of a small percentage of aluminium, is, as I have found, more magnetizable than cast iron, and not far inferior to wrought iron. It should form an excellent material for the cores of electromagnets for many purposes where a cheap manufacture is wanted.

A very useful alternative mode of studying the results obtained by experiment is to construct curves, such as those of Fig. 20, in which the values of the permeability

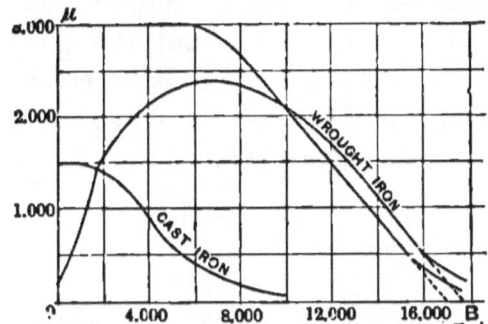

FIG. 20.—CURVES OF PERMEABILITY.

are plotted out vertically in correspondence with the values of **B** plotted horizontally. It will be noticed that in the case of Hopkinson's specimen of annealed wrought iron, between the points where **B** = 7,000 and **B** = 16,000 the mean values of μ lie almost on a straight line, and might be approximately calculated from the equation:

$$\mu = (17,000 - \mathbf{B}) \div 3.5.$$

THE LAW OF THE ELECTROMAGNET.

Many attempts have been made, by Müller, Lamont, Frölich, and others to discover a simple algebraic formula whereby to express the relation between the magnetizing force and the magnetism produced in the electromagnet. According to Müller, these are related to one another in the same proportions as the natural tangent is related to the arc which it subtends. The formulæ of Lamont and Frölich, which are more nearly in keeping with the facts, are based upon the assumption of a relation between the permeability and the degree of magnetization present. Suppose we assume the approximation stated above, that the permeability is proportional to the difference between **B** and some higher limiting value (17,000 for wrought iron, 7,000 for cast iron). If this higher value is called β we may write

$$\mu = \frac{\beta - \mathbf{B}}{a},$$

where a is a constant that varies with the quality of the iron or steel.

Now

$$\mathbf{B} = \mu \mathbf{H};$$

giving by substitution and an easy transformation

$$\mathbf{B} = \beta \frac{\mathbf{H}}{a + \mathbf{H}},$$

which is one form of Frölich's well-known formula. The constant, a, stands for the "diacritical" value of the magnetizing force, or that value which will bring up **B** to half the assumed limiting or "satural" value..

All such formulæ, however convenient, are insufficient, inasmuch as they fail to take into account the properties of the entire magnetic circuit.

HYSTERESIS.

I have already drawn attention to the difference between the ascending and descending curves of magnetization, and may now point out that this is a part of a set of general phenomena of residual effects. The best known of these effects is, of course, the existence in some kinds of iron, and notably in steel, of a remanent or sub-permanent magnetization after the magnetizing

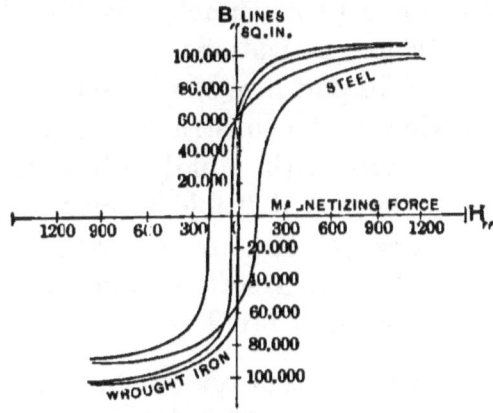

FIG. 21.—CURVES OF HYSTERESIS.

force has been entirely removed. To this retardation of effects behind the causes that produce them the name of "hysteresis" has been given by Prof. Ewing. If a piece of iron is subjected to a magnetizing force which increases to a maximum, then is decreased down to zero, then reversed and carried to a negative maximum, then decreased again to zero, and so carried round an entire cycle of magnetic operations, it is observed that the curves of magnetization form a closed area similar in general to those shown in Fig. 21. This closed area represents the work which has been wasted or dissipated in subjecting the iron to these alternate magnetizing forces. In very soft iron, where the ascending and descending curves are close together, the inclosed area is small, and as a matter of fact very little energy is dissipated in a cycle of magnetic operations. On the other hand, with hard iron, and particularly with steel, there is a great width between the curves and there is a great waste of energy. Hysteresis may be regarded as a sort of internal or molecular magnetic friction, by reason of which alternate magnetizations cause the iron to grow hot. Hence the importance of understanding this curious effect, in view of the construction of electromagnets that are to be used with rapidly alternating currents. The following figures of Table V. give the number of watts (one watt = $\frac{1}{746}$ of a horse power) wasted by hysteresis in well-laminated soft wrought iron when subjected to a succession of rapid cycles of magnetization.

TABLE V.—WASTE OF POWER BY HYSTERESIS.

B	B.	Watts wasted per cubic foot at 10 cycles per second.	Watts wasted per cubic foot at 100 cycles per second.
4,000	25,800	40	400
5,000	32,250	57.5	575
6,000	38,700	75	750
7,000	45,150	92.5	925
8,000	51,600	111	1,110
10,000	64,500	156	1,560
12,000	77,400	206	2,060
14,000	90,300	262	2,620
16,000	103,200	324	3,240
17,000	109,650	394	3,940
18,000	116,100	487	4,870

It will be noted that the waste of energy increases as

the magnetization is pushed higher and higher in a disproportionate degree, the waste when **B** is 18,000 being six times that when **B** is 6,000. In the case of hard iron or of steel the heat waste would be far greater.

Another kind of after-effect was discovered by Ewing, and named by him "viscous hysteresis." This is the name given to the gradual creeping up of the magnetization when a magnetic force is applied with absolute steadiness to a piece of iron. This gradual creeping up may go on for half an hour or more, and amount to several per cent. of the total magnetization.

Another important matter is that all such actions as hammering, rolling, twisting, and the like, impair the magnetic quality of annealed soft iron. Annealed wrought iron which has never been touched by a tool shows hardly any trace of residual magnetization, even after the application of magnetic forces. But the touch of the file will at once spoil it. Sturgeon pointed out the great importance of this point. In the specification for tenders for instruments for the British Postal Telegraphs, it is laid down as a condition to be observed by the constructor that the cores must not be filed after being annealed. The continual hammering of the armature of an electromagnet against the poles may in time produce a similar effect.

FALLACIES AND FACTS ABOUT ELECTROMAGNETS.

I will conclude this lecture by stating a few of the fallacies that are current about electromagnets, and will add to them a few facts, some of which seem paradoxical. The refutation of the fallacies and the explanation of the facts will come in due course.

Fallacies.—The attraction of an electromagnet for its armature varies inversely as the square of its distance from the poles.

The outer windings of an electromagnet are necessarily less effective than those that are close to the iron.

Hollow iron cores are as good as solid cores of the same size.

Pole pieces add to the lifting power of an electromagnet.

It hurts an electromagnet (or, for that matter, a steel magnet) to pull off the keeper suddenly. [It is the sudden slamming on that in reality hurts it.]

The resistance of the coil of an electromagnet ought to be equal to the resistance of the battery.

A coil wound left-handedly magnetizes a magnet differently from a coil wound right-handedly. [It is not a question of winding of coil, but of circulation of current.]

Thick wire electromagnets are less powerful than thin wire electromagnets.

A badly insulated electromagnet is more powerful than one that is well insulated.

A square iron core is less powerful (as Dal Negro says, eighteen-fold!) than a round core of equal weight.

The attraction of an electromagnet for its keeper is necessarily less strong (one-third according to Du Moncel) sidewise than when the keeper is in front of the poles.

Putting a tube of iron outside the coils of an electromagnet makes it attract a distant armature more powerfully.

Facts.—A bar electromagnet with a convex pole holds on tighter to a flat-ended armature than one with a flat pole does.

A thin round disc of iron laid upon the flat round end of an electromagnet (the pole end being slightly larger than the disc), the disc is not attracted, and will not stick on, even if laid down quite centrally.

If a flat armature of iron be presented to the poles of a horseshoe electromagnet the attraction at a short distance is greater, if the armature is presented flankwise, than if it is presented edgewise. On the contrary, the tractive force in contact is greater edgewise than flankwise.

Electromagnets with long limbs are practically no better than those with short limbs for sticking on to masses of iron.

LECTURE II.

GENERAL PRINCIPLES OF DESIGN AND CONSTRUCTION —PRINCIPLE OF THE MAGNETIC CIRCUIT.

To-night we have to discuss the law of the magnetic circuit in its application to the electromagnet, and in particular to dwell upon some experimental results which have been obtained from time to time by different authorities as to the relation between the construction of the various parts of an electromagnet and the effect of that construction on its performance. We have to deal not only with the size, section, length, and material of the iron cores, and of the armatures of iron, but we have to consider also the winding of the copper coil and its form; and we have to speak in particular about the way in which the shaping of the core and of the armature affects the performance of the electromagnet in acting on its armature, whether in contact or at a distance. But before we enter on the last more difficult part of the subject, we will deal solely and exclusively with the law of force of the magnet upon its armature when the two are in contact with one another; in other words, with the law of traction.

I alluded in a historical manner in my first lecture to the principle of the magnetic circuit, telling you how the idea had gradually grown up, perforce, from a consideration of the facts. The law of the magnetic circuit was, however, first thrown into shape in 1873 by Professor Rowland, of Baltimore. He pointed out that if you consider any simple case, and find, as electricians do for the electric circuit, an expression for the magnetizing force which tends to drive the magnetism round the circuit, and divide that by the resistance to magnetization reckoned also all round the circuit, the quotient of those two gives you the total amount of flow or flux of magnetism. That is to say, one may calculate the quantity of magnetism that passes in that way round the magnetic circuit in exactly the same way as one calculates the strength of the electric current by the law of Ohm. Rowland, indeed, went a great deal further than this, for he applied this very calculation to the experiments made by Joule more than 30 years before, and from those experiments deduced the degree of magnetization to which Joule had driven the iron of his magnets, and by inference obtained the amount of current that he had been causing to circulate. Now, this law requires to be written out in a form that can be used for future calculation. To put it in words without any symbols, we must first reckon out from the number of turns of wire in the coil, and the number of ampères of current which circulates in them, the whole *magnetomotive force*—the whole of that which tends to drive magnetism along the piece of iron—for it is, in fact, proportional to the strength of the current and the number of times it circulates. Next we must ascertain the resistance which the magnetic circuit offers to the passage of the magnetic lines. I here avowedly use

Joule's own expression, which was afterward adopted by Rowland, and, for short, so as to avoid having four words, we may simply call it the *magnetic resistance*. Mr. Heaviside has suggested as an advisable alternative term magnetic *reluctance*, in order that we may not confuse the resistance to magnetism in the magnetic circuit with the resistance to the flow of current in an electric circuit. However, we need not quarrel about terms; magnetic reluctance is sufficiently expressive. Then having found these two, the quotient of them gives us a number representing—I must not call it the strength of the magnetic current—I will call it simply the quantity or number of magnetic lines which flow round the circuit; or if we could adopt a term which is used on the continent, we might call it simply *the magnetic flux*, the flux of magnetism being the analogue of the flow of electricity in the electric law. The law of the magnetic circuit may then be stated as follows:

$$\text{Magnetic flux} = \frac{\text{magneto-motive force}}{\text{reluctance.}}$$

However, it is more convenient to deal with these matters in symbols, and therefore the symbols which I use, and have long been using, ought to be explained to you. For the number of spirals in a winding I use the letter S; for the strength of current, or number of ampères, the letter i; for the length of bar, or core, I am going to use the letter l; for the area of cross-section, the letter A; for the permeability of the iron which we discussed in the last lecture, the Greek symbol μ; and for the total magnetic flux, the number of magnetic lines, I use the letter **N**. Then our law becomes as follows:

$$\text{Magneto-motive force } \frac{4\pi \cdot Si}{10};$$

$$\text{Magnetic reluctance } \Sigma \frac{l}{A\mu};$$

$$\text{Magnetic flux} \ldots \ldots \mathbf{N} = \frac{\frac{4\pi \cdot Si}{10}}{\Sigma \frac{l}{A\mu}}.$$

If we take the number of spirals and multiply by the number of ampères of current, so as to get the whole amount of circulation of electric current expressed in so many ampère turns, and multiply by 4π, and divide by 10, in order to get the proper unit (that is to say, multiply it by 1.257), that gives us the magneto-motive force. For magnetic reluctance, calculate out the reluctance exactly as you would the resistance of an electric conductor to the flow of electricity, or the resistance of a conductor of heat to the flow of heat; it will be proportional to the length, inversely proportional to the cross-section, and inversely proportional to the conductivity, or, in the present case, to the magnetic permeability. Now if the circuit is a simple one, we may simply write down here the length, and divide it by the area of the cross-section and the permeability, and so find the value of the reluctance. But if the circuit be not a simple one, if you have not a simple ring of iron of equal section all round, it is necessary to consider the circuit in pieces as you would an electric circuit, ascertaining separately the reluctance of the separate parts, and adding all together. As there may be a number of such terms to be added together, I have prefixed the expression for the magnetic reluctance by the sign Σ of summation. But it does not by any means follow, because we can write a thing down as simply as that, that the calculation of it will be a very simple matter. In the case of magnetic lines we are quite unable to do as one does with electric currents, to insulate the flow. An electric current can be confined (provided we do not put it in at 10,000 volts pressure, or anything much bigger than that) to a copper conductor by an adequate layer of adequately strong—and I use the word "strong" both in a mechanical and electrical sense —of adequately strong insulating material. There are materials whose conductivity for electricity as compared with copper may be regarded perhaps as millions of millions of millions of times less; that is to say, they are practically perfect insulators. There are no such things for magnetism. The most highly insulating substance we know of for magnetism is certainly not 10,000 times less permeable to magnetism than the most highly magnetizable substance we know of, namely, iron in its best condition; and when one deals with electromagnets where curved portions of iron are surrounded with copper, or with air, or other electrically insulating material, one is dealing with substances whose permeability, instead of being infinitely small compared with that of iron, is quite considerable. We have to deal mainly with iron when it has been well magnetized. Its permeability compared with air is then from 1,000 to 100 roughly; that is to say, the permeability of air compared with the iron is not less than from $\frac{1}{100}$th to $\frac{1}{1000}$th part. That means that it is quite possible to have a very considerable leakage of magnetic lines from iron into air occurring to complicate one's calculations and prevent an accurate estimate being made of the true magnetic reluctance of any part of the circuit. Suppose, however, that we have got over all these difficulties and made our calculations of the magnetic reluctance; then dividing the magneto-motive force by the reluctance gives us the whole number of magnetic lines.

There, then, is in its elementary form the law of the magnetic circuit stated exactly as Ohm's law is stated for electric circuits. But, as a general rule, one requires this magnetic law for certain applications, in which the problem is not to calculate from those two quantities what the total of magnetic lines will be. In most of the cases a rule is wanted for the purpose of calculating back. You want to know how to build a magnet so as to give you the requisite number of magnetic lines. You start by assuming that you need to have so many magnetic lines, and you require to know what magnetic reluctance there will be, and how much magneto-motive force will be needed. Well, that is a matter precisely analogous to those which every electrician comes across. He does not always want to use Ohm's law in the way in which it is commonly stated, to calculate the current from the electromotive force and the resistance; he often wants to calculate what is the electromotive force which will send a given current through a known resistance. And so do we. Our main consideration to-night will be devoted to the question how many ampère turns of current circulation must be provided in order to drive the required quantity of magnetism through any given magnetic reluctance. Therefore, we will state our law a little differently. What we want to calculate out is the number of ampère turns required. When once we have got that, it is easy to say what the copper wire must consist of, what sort of wire, and how much of it. Turning then to our algebraic rule, we must transform it, so as to get all the other things besides the ampère turns to the other side of the equation. So we write the formula:

$$S\,i = \frac{\mathbf{N} \cdot \Sigma \frac{l}{A\,\mu}}{1.257}$$

We shall have then the ampère turns equal to the number of magnetic lines we are going to force round the circuit multiplied by the sum of the magnetic reluctances divided by 1.257. Now this number, 1.257, is the constant that comes in when the length l is expressed in centimetres, the area in square centimetres, and the permeability in the usual numbers. Many persons unfortunately—I say so advisedly because of the waste of brain labor that they have been compelled to go through—prefer to work in inches and pounds and feet. They have, in fact, had to learn tables instead of acquiring them naturally without any learning. If the lengths be specified in inches and areas in square inches, then the constant is a little different. The constant in that case, for inch and square inch measures, is 0.3132, so that the formula becomes:

$$Si = \mathbf{N} \times \Sigma \frac{l''}{A'' \mu} \times 0.3132.$$

Here it is convenient to leave the law of the magnetic circuit, and come back to it from time to time as we require. What I want to point out before I go to any of the applications is, that with the guidance provided by this law, one after another the various points that come under review can be arranged and explained, and that there does not now remain—if one applies this law with judgment—a simple fact about electromagnets which is either anomalous or paradoxical. Paradoxical some things may seem in form, but they all reduce to what is perfectly rational when one has a guiding principle of this kind to tell you how much magnetization you will get under given circumstances, or to tell you how much magnetizing power you require in order to get a given quantity of magnetization. I am using the word "magnetization" there in the popular sense, not in the narrow mathematical sense in which it has sometimes been used (*i.e.*, for the magnetic moment per unit cube of the material). I am using it simply to express the fact that the iron or air, or whatever it may be, has been subjected to the process which results in there being magnetic lines of force induced through it.

Now let us apply this law of magnetic circuit in the first place to the traction, that is to say, the lifting power of electromagnets. The law of traction I assumed in my last lecture, for I made it the basis of a method of measuring the amount of permeability. The law of magnetic traction was stated once for all by Maxwell, in his great treatise, and it is as follows:

$$P \text{ (dynes)} = \frac{\mathbf{B}^2 A}{8\pi}$$

Where A is the area in square centimetres this becomes

$$P \text{ (grammes)} = \frac{\mathbf{B}^2 A}{8\pi \times 981}$$

That is, the pull in grammes per square centimetre is equal to the square of the magnetic induction, $\mathbf{B}$ (being the number of magnetic lines to the square centimetre), divided by 8π, and divided also by 981. To bring grammes into pounds you divide by 453.6, so that the formula then becomes:

$$P \text{ (pounds)} = \frac{\mathbf{B}^2 A}{11{,}183{,}000};$$

or if square inch measures are used:

$$P \text{ (pounds)} = \frac{\mathbf{B}_{\prime\prime}^2 A^{\prime\prime}}{72{,}134{,}000}.$$

To save future trouble we will now calculate out from the law of traction the following Table, in which the traction in grammes per square centimetre or in pounds per square inch is set down opposite the corresponding value of $\mathbf{B}$.

TABLE VI.—MAGNETIZATION AND MAGNETIC TRACTION.

B lines per sq. cm.	B. lines per sq. in.	Dynes per sq. centim.	Grammes per sq. centim.	Kilogrs. per sq. centim.	Pounds per sq. inch.
1,000	6,450	39,790	40.56	.0456	.577
2,000	12,900	159,200	162.3	.1623	2.308
3,000	19,350	358,100	365.1	.3651	5.190
4,000	25,800	636,600	648.9	.6489	9.228
5,000	32,250	994,700	1,014	1.014	14.39
6,000	38,700	1,432,000	1,460	1.460	20.75
7,000	45,150	1,950,000	1,987	1.987	28.26
8,000	51,600	2,547,000	2,596	2.596	36.95
9,000	58,050	3,223,000	3,286	3.286	46.72
10,000	64,500	3,979,000	4,056	4.056	57.68
11,000	70,950	4,815,000	4,907	4.907	69.77
12,000	77,400	5,730,000	5,841	5.841	83.07
13,000	83,850	6,725,000	6,855	6.855	97.47
14,000	90,300	7,800,000	7,550	7.550	113.1
15,000	96,750	8,953,000	9,124	9.124	129.7
16,000	103,200	10,170,000	10,390	10.39	147.7
17,000	109,650	11,500,000	11,720	11.72	166.6
18,000	116,100	12,890,000	13,140	13.14	186.8
19,000	122,550	14,630,000	14,650	14.63	208.1
20,000	129,000	15,920,000	16,230	16.23	230.8

This simple statement of the law of traction assumes that the distribution of the magnetic lines is uniform all over the area we are considering; and that unfortunately is not always the case. When the distribution is not uniform then the mean value of the squares becomes greater than the square of the mean value, and consequently the pull of the magnet at its end face may, under certain circumstances, become greater than the calculation would lead you to expect—greater than the average of **B** would lead you to suppose. If the distribution is not uniform over the area of contact then the accurate expression for the tractive force (in dynes) will be

$$P = \frac{1}{\rho\tau} \int \mathbf{B}^2 \, dA,$$

the integration being taken over the whole area of contact.

This law of traction has been verified by experiment. The most conclusive investigations were made about 1886 by Mr. R. H. M. Bosanquet, of Oxford, whose apparatus is depicted in Fig. 22. He took two cores of iron, well faced, and surrounded them both by magnetizing coils, fastened the upper one rigidly, and suspended the other one on a lever with a counterpoise weight. To the lower end of this core he hung a scale-pan, and measured the traction of one upon the other when a known current was circulating a known number of times round the coil. At the same time he placed an exploring coil round the joint, that exploring coil being connected, in the manner with which we were experimenting last week, with a ballistic galvanometer, so that at the moment when the two surfaces parted company, or at the moment when the magnetization was released by stopping the magnetizing current, the galvanometer indication enabled him to say exactly how many magnetic lines went

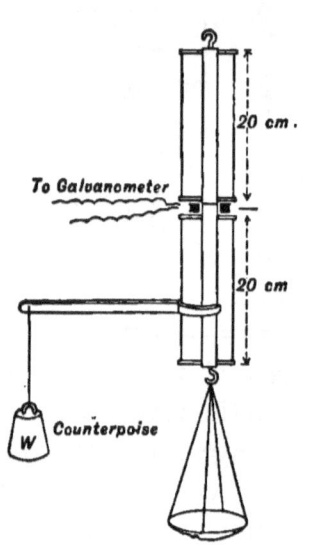

FIG. 22.—BOSANQUET'S VERIFICATION OF THE LAW OF TRACTION.

through that exploring coil. So that, knowing the area, you could calculate the number per square centimetre, and you could therefore compare B^2 with the pull per square centimetre obtained directly on the scale-pan. Bosanquet found that even when the surfaces were not absolutely perfectly faced the correspondence was very close indeed, not varying by more than one or two per cent. except with small magnetizing forces, say forces less than five C. G. S. units.

When one knows how irregular the behavior of iron is when the magnetizing forces are so small as this, one is not astonished to find a lack of proportionality. The correspondence was, however, sufficiently exact to say that the experiments verified the law of traction, that the pull is proportional to the square of the magnetic induction through the area integrated over that area.

Now the law of traction being in that way established, one at once begins to get some light upon the subject of the design of electromagnets. Indeed, without going into any mathematics, Joule had foreseen this when he in some instinctive sort of way seemed to consider that the proper way to regard an electromagnet for the purpose of traction was to think how many square inches of contact surface it had. He found that he could magnetize iron up until it pulled with a force of 175 pounds to the square inch, and he doubted whether a traction as great as 200 pounds per square inch could be obtained.

In the following Table Joule's results (see Table I.) are recalculated, and the values of B deduced:

TABLE VII.—JOULE'S RESULTS RE-CALCULATED.

Description of Electromagnet.	Section.		Load.		Pounds per sq. inch.	Kilos. per sq. cm.	B	Ratio of load to weight.
	sq. in.	sq. cm.	lbs.	kilos.				
Joule's own electro-magnets.. No. 1	10	64.5	209.0	947	104.5	7.35	13,600	139
No. 2	0.196	1.26	49	22	125	8.75	14,700	324
No. 3	0.0436	6.28	12	5.4	137.5	9.75	15,410	1.286
No. 4	0.0012	0.0077	0.202	0.09	81	5.7	11,830	2,384
Nesbit's........	4.5	29.1	142.8	647	158.5	11.2	16,550	28
Henry's........	3.94	25.3	750	346	95	6.7	12,820	36
Sturgeon's	0.196	1.26	53	22.6	127.5	8.95	14,850	114

I will now return to the data in Table VI., and will ask you to compare the last column with the first. Here are various values of **B**, that is to say, the amounts of magnetization you get into the iron. You cannot conveniently crowd more than 20,000 magnetic lines through the square centimetre of the best iron, and, as a reference to the curves of magnetization shows, it is not expedient in the practical design of electromagnets to attempt, except in extraordinary cases, to crowd more than about 16,000 magnetic lines into the square centimetre. The simple reason is this: that if you are working up the magnetic force, say from 0 up to 50, a magnetizing force of 50 applied to good wrought iron will give you only 16,000 lines to the square centimetre, and the permeability by that time has fallen to about 320. If you try to force the magnetization any further, you find that you have to pay for it too heavily. If you want to force another 1,000 lines through the square centimetre, to go from 16,000 to 17,000, you have to add on an enormous magnetizing force; you have to double the whole force from that point to get another 1,000 lines added. Obviously it would be much better to take a larger piece of iron and not to magnetize it too highly —to take a piece a quarter as large again, and to magnetize that less forcibly. It does not therefore pay to go much above 16,000 lines to a square centimetre—that is to say, expressing it in terms of the law of traction, and the pounds per square inch, it does not pay to design your electromagnet so that it shall have to carry more than about 150 pounds to the square inch. This shall be our practical rule: let us at once take an example. If you want to design an electromagnet to carry a load of one ton, divide the ton, of 2,240 pounds, by 150, and that gives the requisite number of square inches of wrought iron, namely, 14.92, or say 15. Of course one would work with a horseshoe shaped magnet, or something equivalent—something with a return circuit—and calculate out the requisite cross-section, so that the total area exposed might be sufficient to carry the given load at 150 pounds to the square inch. And, as a horseshoe magnet has two poles, the cross-section of the bar of which it is made must be $7\frac{1}{2}$ square inches. If of round iron, it must be about $3\frac{1}{2}$ inches in diameter; if of square iron, it must be $2\frac{3}{4}$ inches each way.

That settles the size of the iron, but not the length. Now, the length of the iron, if one only considers the law of the magnetic circuit, ought to be as short as it can possibly be made. Reflect for what purpose we are designing. The design of an electromagnet is to be considered, as every design ought to be, with a view to the ultimate purpose to be served by that which you are designing. The present purpose is the actual sticking on of the magnet to a heavy weight, not acting on another magnet at a distance, not pulling at an armature separated from it by a thick layer of air; we are dealing with traction in contact. The question is, How long a piece of iron shall we need to bend over? The answer is: Take length enough, and no more than enough, to permit of room for winding on the necessary quantity of wire to carry the current which will give the requisite magnetizing power. But this latter we do not yet know; it has to be calculated out by the law of the magnetic circuit. That is to say, we must calculate the magnetic flux, and the magnetic reluctance as best we can; then from these calculate the ampère turns of current; and from this calculate the needful quantity of copper wire, so arriving finally at the proper length of the iron core. It is obvious the cross-section being given and the value of **B** being prescribed, that settles the whole number of magnetic lines, **N**, that will go through the section. It is self-evident that length adds to the magnetic reluctance, and, therefore, the longer the length is, the greater have to be the number of ampère turns of circulation of the current; while the less the length is, the smaller need be the number of ampère turns of circulation. Therefore you should design the electromagnet as stumpy as possible, that is to say make it a stumpy arch, even as Joule did when he came across the same problem, and arrived, by a sort of scientific instinct, at the right solution. You should have no greater length of iron than is necessary in order to get the windings on. Then you see we cannot absolutely calculate the length of the iron until we have an idea about the winding, and we must settle, therefore, provisionally, about the windings. Take a simple ideal case. Suppose we had an indefinitely long, straight iron rod, and we wound that from end to end with a magnetizing coil. How thick a coil, how many ampère turns of circulation per inch length will you require in order to magnetize up to any particular degree? It is a matter of very simple calculation. You can calculate exactly what the magnetic reluctance of an inch length of the core will be. For example, if you are going to magnetize up to 16,000 lines per square centimetre, the permeability will be 320. You can take the area anything you like, and consider the length of one inch; you can therefore calculate the magnetic reluctance per inch of conductor, and then you can at once say how many ampère turns per inch would be necessary in order to give the desired indication of 16,000 magnetic lines to the square centimetre. And knowing the properties of copper wire, and how it heats up when there is a current; and knowing also how much heat you can get rid of per square inch of surface, it is a very simple matter to calculate what minimum thickness of copper the fire insurance companies would allow you to use. They would not allow you to have too thin a copper wire, because if you provide an insufficient thickness of copper you still must drive your ampères through it to get a sufficient number of ampère turns per inch of length; and if you drive those ampères through copper winding of an insufficient thickness the copper wire will overheat and your insurance policy will be revoked. You therefore are compelled, by the practical consideration of not overheating, to provide a certain thickness of copper wire winding. I have made a rough calculation for certain cases, and I find that for such small electromagnets as one may ordinarily deal with, it is not necessary in any practical case to use a copper wire winding, the total thickness of which is greater than about half an inch; and, as a matter of fact, if you use as much thickness as half an inch, you need not then wind the coil all along, for if you will use copper wire winding, no matter what the size, whether thin or thick, so that the total thickness of copper outside the iron is half an inch, you can without overheating, using good wrought iron, make one inch of winding do for 20 inches length of iron. That is to say, you do not really want more than $\frac{1}{40}$th of an inch of thickness of copper outside the iron to magnetize up to the prescribed degree of saturation that indefinitely long piece of which we are thinking, without overheating the outside surface in such a way as to violate the insurance rules. Take it approximately, if you wind to a thickness of half an inch the inch length of copper will magnetize 20 inches length of iron up to the point where **B** equals 16,000. If then we have a bar bent into a sort of horseshoe in order to make it stick on to a perfectly fitting armature also of equal section and quality, we really do not want more than one inch along the inner curve for every 20 inches of iron. An extremely stumpy magnet, such as I have sketched in Fig. 23, will therefore do, if one can only get the iron sufficiently homogeneous throughout. If, instead of crowding the wire near the polar parts, we could wind entirely all round the curved part, though the layer of copper winding would be half an inch thick inside the arch, it would be much less outside. Such a magnet, provided the armature fitted with perfect accuracy to the polar surfaces, and provided a battery were arranged to send the requisite number of ampères of current through the coils, would pull with a force of one ton, the iron being but 3½ inches in diameter. For my own part, in this case I should prefer not

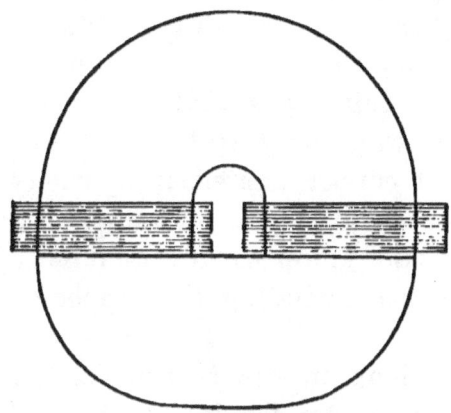

Fig. 23.—Stumpy Electromagnet.

to use round iron, one of square or rectangular section being more convenient; but the round iron would take less copper in winding, as each turn would be of minimum length if the section were circular.

Now, this sort of calculation requires to be greatly modified directly one begins to deal with any other case. A stumpy short magnetic circuit with great cross-section is clearly the right thing for the greatest traction. You will get the given magnetization and traction with the least amount of magnetizing force when you have the area as great as possible, and the length as small as possible. You will kindly note that I have given you as yet no proofs for the practical rules that I have been using; they must come later. Also I have said nothing about the size of the wire, whether thick or thin. That does not in the least matter, for the ampère turns of magnetizing power can be made up in any desired way. Suppose we want on any magnet 100 ampère turns of magnetizing power, and we choose to employ a thin wire that will only carry half an ampère, then we must wind 200 turns of that thin wire. Or, suppose we choose to wind it with a thick wire that will carry 10 ampères, then we shall want only 10 turns of that wire. The same weight of copper, heated up by the corresponding current to an equal degree of temperature, will have equal magnetizing power when wound on the same core. But the rules about winding the copper will be considered later.

Now if you look in the text-books that have been written on magnetism for information about the so-called lifting power or portative force of magnets—in other words, the traction—you will find that from the time of Bernoulli downward, the law of portative force has claimed the attention of experimenters, who, one after another, have tried to give the law of portative force in terms of the weight of the magnets; usually dealing with permanent magnets, not electromagnets. Bernoulli gave[1] a rule something of the following kind, which is commonly known as Häcker's rule:

$$P = a \sqrt[3]{W};$$

---

[1] *Acta Helvetica*, III., p. 233, 1758,

where W is the weight of the magnet, P the greatest load it will sustain, and a a constant depending on the unit of weight chosen, on the quality of the steel and on its goodness of magnetization. If the weights are in pounds, then a is found for the best steels to vary from 18 to 24 in magnets of horseshoe shape. This expression is equivalent to saying that the power which a magnet can exert—he was dealing with steel magnets; there were no electromagnets in Bernoulli's time—is equal to some constant multiplied by the three-halfth root of the weight of the magnet itself. The rule is accurate only if you are dealing with a number of magnets all of the same geometrical form, all horseshoes, let us say, of the same general shape, made from the same sort of steel, similarly magnetized. In former years I pondered much on Häcker's rule, wondering how on earth the three-halfth root of the weight could have anything to do with the magnetic pull; and, having cudgeled my brains for a considerable time, I saw that there was really a very simple meaning in it. What I arrived at [2] was this: If you are dealing with a given material, say hard steel, the weight is proportional to the volume, and the cube root of the volume is something proportional to the length, and the square of the cube root forms something proportional to the square of the length, that is to say, to something of the nature of a surface. What surface? Of course the polar surface. This complex rule when thus analyzed turns out to be merely a mathematician's expression of the fact that the pull for a given material magnetized in a given

---

[2] *Philosophical Magazine*, July, 1888.

way is proportional to the area of the polar surface; a law which in its simple form Joule seems to have arrived at naturally, and which in this extraordinarily academic form was arrived at by comparing the weights of magnets with the weight which they would lift. You will find it stated in many books that a good magnet will lift 20 times its own weight. There never was a more fallacious rule written. It is perfectly true that a good steel horseshoe magnet weighing one pound ought to be able to pull with a pull of 20 pounds on a properly shaped armature. But it does not follow that a magnet which weighs two pounds will be able to pull with a force of 40 pounds. It ought not to, because a magnet that weighs two pounds has not poles twice as big if it is the same shape. In order to have poles twice as big you must remember that three-halfth root coming in. If you take a magnet that weighs eight times as much, it will have twice the linear dimensions and four times the surface; and with four times the surface in a magnet of the same form, similarly magnetized, you will have four times the pull. With a magnet eight times as heavy you will have only four times the pull. The pull, when other things are equal, goes by surface and not by weight, and therefore it is ridiculous to give a rule saying how many times its own weight a magnet will pull. It is also narrated as a very extraordinary thing that Sir Isaac Newton had a magnet, a loadstone, which he wore in a signet ring, which would lift 234 times its own weight. I have had an electromagnet which would lift 2,500 times its own weight, but then it was a very small one, and did not weigh more than a grain and a half. When you come to small things, of course the surface is large proportionally to the weight; the smaller you go, the larger becomes that disproportion. This all shows that the old law of traction in that form was practically valueless, and did not guide you to anything at all, whereas the law of traction as stated by Maxwell, and explained further by the law of the magnetic circuit, proves a most useful rule.

From this digression let us return to the law of the magnetic circuit. I gave you in my first lecture, when speaking of permeability, the following rule for calculating the magnetic induction **B**: Take the pull in pounds, and the area of cross-section in square inches; divide one by the other, and take the square root of the quotient; then multiplying by 1,317 gives **B**; or multiplying by 8,494 gives $\mathbf{B}_u$. We have therefore a means of stepping from the pull per square inch to $\mathbf{B}_u$, or from $\mathbf{B}_u$ to the pull per square inch. Now the other rule of the magnetic circuit also enables us to get from the ampère turns down to $\mathbf{B}_u$, for we have the following expression for the ampère turns:

$$Si = \mathbf{N} \times \Sigma \frac{l''}{A'' \mu} \times 0.3132,$$

and **N**, the whole number of magnetic lines in the magnetic circuit, is equal to $\mathbf{B}_u$ multiplied by A'', or

$$\mathbf{N} = \mathbf{B}_u A''.$$

From these we can deduce a simple direct expression, provided we assume the quality of iron as before, and also assume that there is no magnetic leakage, and that the area of cross-section is the same all round the circuit, in the armature as well as in the magnet core. So that l'' is simply the mean total path of the magnetic lines all round the closed magnetic circuit. We may then write:

$$S i = \frac{\mathbf{B}_u\, l''}{\mu} \times 0.3132;$$

whence

$$\mathbf{B}_u = \frac{\mu \times S i}{l'' \times 0.3132}.$$

But by the law of traction, as stated above,

$$\mathbf{B}_u = 8{,}494\ \sqrt{\frac{P\,(\text{lbs.})}{A\,(\text{sq. in.})}}$$

Equating together these two values of $\mathbf{B}_u$, and solving, we get for the requisite number of ampère turns of circulation of exciting currents:

$$S i = 2{,}661 \times \frac{l''}{\mu} \times \sqrt{\frac{P\,(\text{lbs.})}{A\,(\text{sq. in.})}}$$

This, put into words, amounts to the following rule for calculating the amount of exciting power that is required for an electromagnet pulling at its armature, in the case where there is a closed magnetic circuit with no leakage of magnetic lines. Take the square root of the pounds per square inch; multiply this by the mean total length (in inches) all round the iron circuit; divide by the permeability (which must be calculated from the pounds per square inch by help of Table VI. and Table II.), and finally multiply by 2,661; the number so obtained will be the number of ampère turns. One goes then at once from the pull per square inch to the number of ampère turns required to produce that pull in a magnet of given length and of the prescribed quality. In the case where the pull is specified in kilogrammes, the area of section in square centimetres, and the length in centimetres, the formula becomes

$$S i = 3{,}951 \cdot \frac{l}{\mu} \sqrt{\frac{P}{A}}.$$

As an example, take a magnet core of round annealed wrought iron, half an inch in diameter, eight inches long, bent to horseshoe shape. As an armature, another piece, four inches long, bent to meet the former. Let us agree to magnetize the iron up to the pitch of pulling with 112 pounds to the square inch. Reference to Table VI. shows that B_u will be about 90,000, and Table II. shows that in that case μ will be about 907. From these data calculate what load the magnet will carry, and how many ampère turns of circulation of current will be needed.

Ans.—Load (on two poles) = 43.97 lbs.
Ampère turns needed = 372.5

N. B.—In this calculation it is assumed that the contact surface between armature and magnet is perfect. It never is; the joint increases the reluctance of the magnetic circuit, and there will be some leakage. It will be shown later how to estimate these effects, and to allow for them in the calculations.

Here let me go to a matter which has been one of the paradoxes of the past. In spite of Joule, and of the laws of traction, showing that the pull is proportional to the area, you have this anomaly—that if you take a bar magnet having flat-ended poles, and measure the pull which its pole can exert on a perfectly flat armature, and then deliberately spoil the truth of the contact surface, rounding it off, so making the surface gently convex, the convex pole, which only touches at a portion of its area instead of over the whole, will be found to exert a bigger pull than the perfectly flat one. It has been shown by various experimenters, particularly by Nicklès, that if you want to increase the pull of a magnet with armatures you may reduce the polar surface. Old steel magnets were frequently purposely made with a rounded contact surface. There are plenty of examples. Suppose you take a straight round core, or one leg of a horseshoe, which answers equally, and take a flat-ended rod of iron of the same diameter as an armature; stick it on endwise, and measure the pull when a given amount of ampère turns of current is circulating round. Then, having measured the pull, remove it and file it a little, so as to reduce it at the edges, or take a slightly narrower piece of iron, so that it will actually be exerting its power over a smaller area, you will get a greater pull. What is the explanation of this extraordinary fact? A fact it is, and I will show it to you. Here, Fig. 24, is a small electromagnet which we can place with its poles upward. This was very carefully made, the iron poles very nicely faced, and on coming to try them it was found they were nearly equal, but one pole, A, was a little stronger than the other. We have, therefore, rounded the other pole, B, a little, and here I will take a piece of iron, C, which has itself been slighty rounded at one end, though it is flat at the other. I now turn on the current to the electromagnet, and I take a spring balance so that we can measure the pull at either of the two poles. When I put the flat end of C to the flat pole A so that there is an excellent contact, I find the pull about $2\frac{1}{2}$ pounds. Now try the round end of C on the flat pole A; the pull is about three pounds. The flat end of C on the round pole B is also about three pounds. But if now I put together two surfaces that are both rounded I get almost exactly the same pull as at first with the two flat surfaces. I

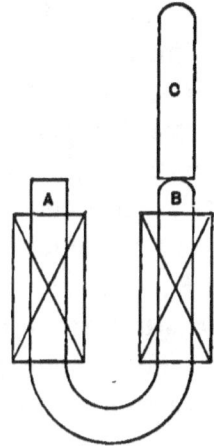

Fig. 24.—Experiment on Rounding Ends.

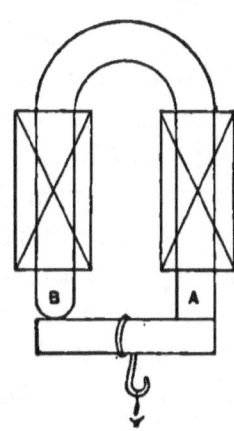

Fig. 25.—Experiment of Detaching Armature.

have made many experiments on this, and so have others. Take the following case: There is hung up a horseshoe magnet, one pole being slightly convex and the other absolutely flattened, and there is put at the bottom a square bar armature, over which is slipped a hook to which weights can be hung. Which end of the armature do you think will be detached first?

If you were going simply by the square inches, you would say this square end will stick on tighter; it has more gripping surface. But, as a matter of fact, the other sticks tighter. Why? We are dealing here with a magnetic circuit. There is a certain total magnetic reluctance all round it, and the whole number of magnetic lines generated in the circuit depends on two things—on the magnetizing force, and on the reluctance all round; and, saving a little leakage, it is the same number of magnetic lines which come through at B as go through at A. But here, owing to the fact that there is at B a better contact at the middle than at the edges of the pole, the lines are crowded into a smaller space, and therefore at that particular place B_u the number of lines per square inch runs up higher, and when you square the larger number, its square becomes still larger in proportion. In comparing the square of smaller B_u with the square of greater B_u, the square of the smaller B_u over the larger area turns out to be less than the square of the larger B_u integrated over the smaller area. It is the law of the square coming in.

As an example, take the case of a magnet pole formed on the end of a piece of round iron 1.15 inches in diameter. The flat pole will have 1.05 inches area. Suppose the magnetizing forces are such as to make $B_u = 90,300$, then by Table VI. the whole pull will be 118.75 pounds, and the actual number of lines through the contact surface will be $N = 94,815$. Now suppose the pole be reduced by rounding off the edge till the effective contact area is reduced to 0.9 square inch. If all these lines were crowded through that area, that would give a rate of 105,350 per square inch. Suppose, however, that the additional reluctance and the leakage reduced the number by two per cent., there would still be 103,260 per square inch. Reference to Table VI. shows that this gives a pull of 147.7 pounds per square inch, which, multiplied by the reduced area 0.9, gives a total pull of 132.9 pounds, which is larger than the original pull.

Let me show you yet another experiment. This is the same electromagnet (Fig. 24) which has one flat pole and one rounded pole. Here is an armature, also bent, having one flat and one rounded pole. If I put flat to flat and round to round, and pull at the middle,

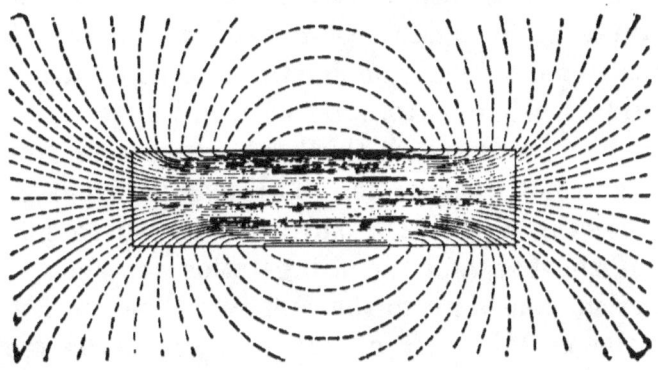

Fig. 26.—Lines of Force Running through Bar Magnet.

the flat to flat detaches first; but if we take round to flat and flat to round, we shall probably find they are about equally good—it is hard to say which holds the stronger.

The law of traction can again be applied to test the so-called distribution of free magnetism on the surface. This is a subject on which I shall have to say a good deal. We must therefore carefully consider what is meant by the phrase. Let Fig. 26 be a rough drawing of an ordinary bar magnet. Every one knows that if we dip such a magnet into iron filings the small bits of iron stick on more especially at the ends, but not exclusively, and if you hold it under a piece of paper or cardboard, and sprinkle iron filings on the paper, you obtain curves like those shown on the diagram. They attest the distribution of the magnetic forces in the external space. The magnetism running internally through the body of the iron begins to leak out sidewise, and, finally, all the rest leaks out in a great tuft at the end. These magnetic lines pass round to the other end and there go in again. The place where the steel is internally most highly magnetized is this place across the middle, where externally no iron filings at all stick to it. Now, we have to think of magnetism from the inside and not the outside. This magnetism extends in lines, coming up to the surface somewhere near the ends of the bar, and the filings stick on wherever the magnetism comes up to the surface. They do not stick on at the middle part of the bar, where the metal is really most completely permeated through and through by the magnetism; there are a larger number of lines per square centimetre of cross-section in the middle region where none come up to the surface, and no filings stick on. Now, we may explore the leakage of magnetic lines at various points of the surface of the magnet by the method of traction. We can thereby arrive at a kind of measure of the amount of magnetism that is leaking, or, if you like to call it so, of the intensity of the "free magnetism" at the surface. I do not like to have to use these ancient terms, because they suggest the ancient notion that magnetism was a fluid or, rather, two fluids, one of which was plastered on at one end of the magnet, and the other at the other, just as you might put red paint or blue paint over the ends. I only use that term because it is already more or less familiar. Here is one of the ways of experimentally exploring the so-called distribution of free magnetism. The method was, I believe, originally due to Plücker; at any rate, it was much used by him. This little piece of apparatus was arranged by my friend and predecessor, Prof. Ayrton, for the purpose of teaching his students at the Finsbury College.[3] Here is a bar magnet of steel, marked in centimetres from end to end; over the top of it there is a little steel-yard, consisting of a weight sliding along an arm. At the end of that steel-yard there is suspended a small bullet of iron. If we bring that bullet into contact with the bar magnet anywhere near the end, and equilibrate the pull by sliding the counterpoise along the steel-yard arm, we shall obtain the definite pull required to detach that piece of iron. The pull will be proportional, by Maxwell's rule, to the square of the number of magnetic lines coming up from the bar into it. Shift the magnet on a whole centimetre, and attach the bullet a little further on; now equilibrate it, and we shall find it will require a rather smaller force to detach it. Try it again, at points along from the end to the middle. The greatest force required to detach it will be found at the extreme corner, and a little less a little way on, and so on until we find at the middle the bullet does not stick on at all, simply because there are here no magnetic lines leaking. The method is not perfect, because it obviously depends

---

[3] See Ayrton's "Practical Electricity," Fig. 5a, p. 24.

on the magnetic properties of the little bullet, and whether it is much or little saturated with magnetism. Moreover, the presence of the bullet perturbs the very thing that is to be measured. Leakage into air is one thing; leakage into air perturbed by the presence of the little bullet of iron, which invites leakage into itself, is another thing. It is an imperfect experiment at the best, but a very instructive one. This method has been used again and again in various cases for exploring the apparent magnetism on the surface. I shall use it hereafter, reserving the right to interpret the result by the light of the law of traction.

I now pass to the consideration of the attraction of a magnet on a piece of iron at a distance. And here I come to a very delicate and complicated question. What is the law of force of a magnet—or electromagnet—acting at a point some distance away from it? I have a very great controversy to wage against the common way of regarding this. The usual thing that is proper to say is that it all depends on the law of inverse squares. Now, the law of inverse squares is one of those detestable things needing to be abolished, which, although it may be true in abstract mathematics, is absolutely inapplicable with respect to electromagnets. The only use, in fact, of the law of inverse squares, with respect to electromagnetism, is to enable you to write an answer when you want to pass an academical examination, set by some fossil examiner, who learned it years ago at the University, and never tried an experiment in his life to see if it was applicable to an electromagnet. In academical examinations they always expect you to give the law of inverse squares. What is the law of inverse squares? We had better understand what it is before we condemn it. It is a statement to the following effect —that the action of the magnet (or of the pole, some people say), at a point at a distance away from it, varies inversely as the square of the distance from the pole. There is a certain action at one inch away. Double the distance; the square of that will be four, and, inversely, the action will be one-quarter; at double the distance the action is one-quarter; at three times the distance the action is one-ninth, and so on. You just try it with any electromagnet; nay, take any magnet you like, and unless you hit upon the particular case, I believe you will find it to be universally untrue. Experiment does not prove it. Coulomb, who was supposed to establish the law of inverse squares by means of the torsion balance, was working with long, thin needles of specially hard steel, carefully magnetized, so that the only leakage of magnetism from the magnet might be as nearly as possible leakage in radiating tufts at the very ends. He practically had point poles. When the only surface magnetism is at the end faces, the magnetic lines leak out like rays from a centre, in radial lines. Now the law of inverse squares is never true except for the action of points; it is a *point* law. If you could get an electromagnet or a magnet with poles so small in proportion to its length that you can consider the end face of it as the only place through which magnetic lines leak up into the air, and the ends themselves so small as to be relatively mere points; if, also, you can regard those end faces as something so far away from whatever they are going to act upon that the distance between them shall be large compared with their size, and the end itself so small as to be a point, then, and then only, is the law of inverse squares true. It is a law of the action of points. What do we find with electromagnets? We are dealing with pieces of iron which are not infinitely long with respect to their cross-section, and generally possessing round or square end faces of definite magnitude, which are quite close to the armature, and which are not so infinitely far away that you can consider the polar face a point as compared with its distance away from the object upon which it is to act. Moreover, with real electromagnets there is always lateral leakage; the magnetic lines do not all emerge from the iron through the end face. Therefore, the law of inverse squares is not applicable to that case. What do we mean by a pole, in the first place? We must settle that before we can even begin to apply any law of inverse squares. When leakage occurs all over a great region, as shown in this diagram, every portion of the region is polar; the word polar simply means that you have a place somewhere on the surface of the magnet where filings will stick on; and if filings will stick on to a considerable way down toward the middle, all that region must be considered polar, though more strongly at some parts than at others. There are some cases where you can say that the polar distribution is such that the magnetism leaking through the surface acts as if there were a magnetic centre of gravity a little way down, not actually at the end; but cases where you can say there is such a distribution as to have a magnetic centre of gravity are strictly few. When Gauss had to make up his magnetic measurements of the earth, to describe the earth's magnetism, he found it absolutely impossible to assign any definite centre of gravity to the observed distribution of magnetism over the northern regions of the earth; that, indeed, there was not in this sense any definite magnetic pole to the earth at all. Nor is there to our magnets. There is a

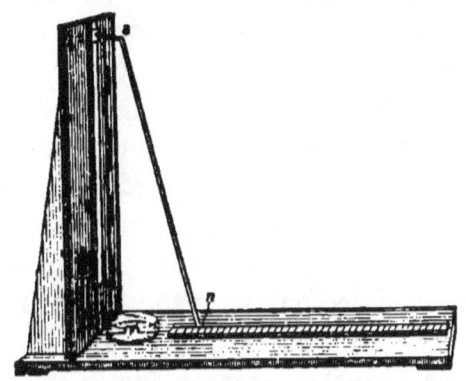

FIG. 27.—APPARATUS TO ILLUSTRATE THE LAW OF INVERSE SQUARES.

polar region, but not a pole; and if there is no centre of gravity of the surface magnetism that you can call a pole from which to measure distance, how about the law of inverse squares? Allow me to show you an apparatus (Fig. 27), the only one I ever heard of in which the law of inverse squares is true. Here is a very long, thin magnet of steel, about three feet long, very carefully magnetized so as to have no leakage until quite close up to the end. The consequence is that for practical purposes you may treat this as a magnet having point poles, about an inch away from the ends. The south pole is upward and the north pole is below, resting in a groove in a base-board which is graduated with a scale, and is set in a direction east and west. I use a long magnet, and keep the south pole well away, so that it shall not perturb the action of the north pole, which, being small, I ask to be allowed to consider as a point. I am going to consider this point as acting on a small compass needle suspended over a card under this glass case, constituting a little magnetometer. If this were properly arranged in a room free from all other magnets, and set so that that needle shall point north, what will be the effect of having the north pole of the long magnet at some distance eastward? It will repel the north end and attract the south, producing a certain deflection which can be read off; reckoning the force which causes it by calculating the tangent of the angle of the deflection. Now, let us move the north pole (regarded as a point) nearer or farther, and study the effect. Suppose we halve the distance from the pole to the indicating needle, the deflecting force at half the distance is four times as great; the force at double the distance is one-quarter as great. Wherefore? Because, firstly, we have taken a case where the distance apart is very great, compared with the size of the pole; secondly, the pole is practically concentrated at a point; thirdly, there is only one pole acting; and fourthly, this magnet is of hard steel, and its magnetism in no way depends on the thing it is acting on, but is constant. I have carefully made such arrangements that the other pole shall be in the axis of rotation, so that its action on the needle shall have no horizontal component. The apparatus is so arranged that, whatever the position of that north pole, the south pole, which merely slides perpendicularly up and down on a guide, is vertically over the needle, and therefore does not tend to turn it round in any direction whatever. With this apparatus one can approximately verify the law of inverse squares. But this is not like any electromagnet ever used for any useful purpose. You do not make electromagnets long and thin, with point poles a very large distance away from the place where they are to act; no, you use them with large surfaces close up to their armature.

There is yet another case which follows a law that is not a law of inverse squares. Suppose you take a bar magnet, not too long, and approach it broadside on toward a small compass needle, Fig. 28. Of course, you know as soon as you get

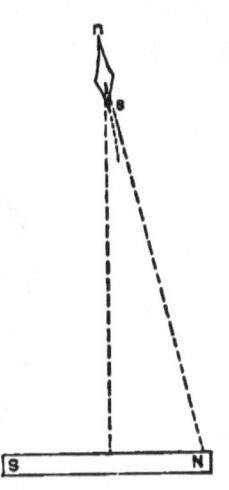

FIG. 28.—DEFLECTION OF NEEDLE CAUSED BY BAR MAGNET BROADSIDE ON.

anywhere near the compass needle it turns round. Did you ever try whether the effect is inversely proportional to the square of the distance reckoned from the middle of the compass needle to the middle of the magnet? Do you think that the deflections will vary inversely with the squares of the distances? You will find they do not. When you place the bar magnet like that, broadside on to the needle, the deflections vary as the cube of the distance, not the square.

Now, in the case of an electromagnet pulling at its armature at a distance, it is utterly impossible to state the law in that misleading way. The pull of the electromagnet on its armature is not proportional to the distance, nor to the square of the distance, nor to the cube, nor to the fourth power, nor to the square root, nor to the three-halfth root, nor to any other power of

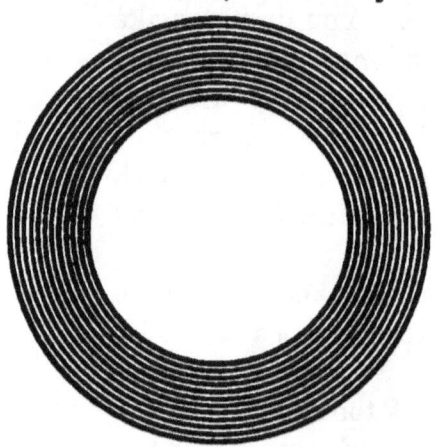

FIG. 29.—CLOSED MAGNETIC CIRCUIT.

the distance whatever, direct or inverse, because you find, as a matter of fact, that as the distance alters something else alters too. If your poles were always of the same strength, if they did not act on one another, if they were not affected by the distance in between, then some such law might be stated. If we could always say, as we used to say in the old language, "at that pole," or "at that point," there are to be considered so many "units of magnetism," and at that other place so many units, and those are going to act on one another; then you could, if you wished, calculate the force by the law of inverse squares. But that does not correspond to anything in fact, because the poles are not points, and further, the quantity of magnetism on them is not a fixed quantity. As soon as the iron armature is brought near the pole of the electromagnet there is a mutual interaction; more magnetic lines flow out from the pole than before, because it is easier for magnetic

FIG. 30.—DIVIDED MAGNETIC CIRCUIT.

lines to flow through iron than through air. Let us consider a little more narrowly that which happens when a layer of air is introduced into the magnetic circuit of an electromagnet. Here we have (Fig. 29) a closed magnetic circuit, a ring of iron, uncut, such as we experimented on last week. The only reluctance in the path of the magnetic lines is that of the iron, and this reluctance we know to be small. Compare Fig. 29 with Fig. 30, which represents a divided ring with air-gaps in between the severed ends. Now, air is a less permeable medium for magnetic lines than iron is, or, in other words, it offers a greater magnetic reluctance. The magnetic permeability of iron varies, as we know, both with its quality and with the degree of magnetic saturation. Reference to Table III. shows that if the iron has been magnetized up so as to carry 16,000 magnetic lines per square centimetre, the permeability at that stage is about 320. Iron at that stage conducts magnetic lines 320 times better than air does; or air offers 320 times as much reluctance to magnetic lines as iron (at that stage) does. So then the reluctance in the gaps to magnetization is 320 times as great as it would have been if the gaps had been filled up with iron. Therefore, if you have the same magnetizing coil with the same battery at work, the introduction of air-gaps into the magnetic circuit will, as a first effect, have the result of decreasing the number of magnetic lines that flow round the circuit. But this first effect itself produces a second effect. There are fewer magnetic lines going through the iron. Consequently if there were 16,000 lines per square centimetre before, there will now be fewer—say only 12,000 or so. Now refer back to Table III. and you will find that when **B** is 12,000 the permeability of the iron is not 320, but 1,400 or so. That is to say, at this stage, when the magnetization of the iron has not been pushed so far, the magnetic reluctance of air is 1,400 times greater than that of iron, so that there is a still greater relative throttling of the magnetic circuit by the reluctance so offered by the air-gaps.

LECTURES ON THE ELECTROMAGNET. 119

Apply that to the case of an actual electromagnet. Here is a diagram, Fig. 31, representing a horseshoe electromagnet with an armature of equal section in contact with it. The actual electromagnet for the experiment is here on the table. You can calculate out from the section, the length of iron and the table of permea-

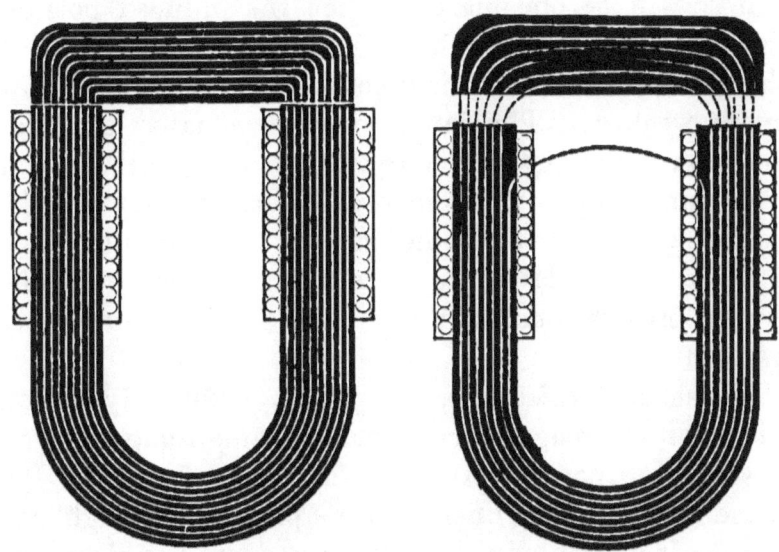

FIG. 31.—ELECTROMAGNET WITH ARMATURE IN CONTACT.

FIG. 32.—ELECTROMAGNET WITH AIR-GAPS ONE MILLIMETRE WIDE.

bility how many ampère turns of excitation will produce any required pull. But now consider that same electromagnet, as in Fig. 32, with a small air-gap between the armature and the polar faces. The same circulation of current will not now give you as much magnetism as before, because you have interposed air-gaps, and by the very fact of putting in reluctance there the number of magnetic lines is reduced.

Try, if you like, to interpret this in the old way by the old notion of poles. The electromagnet has two poles, and these excite induced poles in the opposite surface of the armature, resulting in attraction. If you double the distance from the pole to the iron, the magnetic force (always supposing the poles are mere points) will be one-quarter, hence the induced pole on the armature will only be one-quarter as strong. But the pole of the electromagnet is itself weaker. How much weaker? The law of inverse squares does not give you the slightest clue to this all-important fact. If you cannot say how much weaker the primary pole is, neither can you say how much weaker the induced pole will be, for the latter depends upon the former. The law of inverse squares in a case like this is absolutely misleading.

Moreover, a third effect comes in. Not only do you cut down the magnetism by making an air-gap, but you have a new consideration to take into account. Because the magnetic lines, as they pass up through one of the air-gaps, along the armature, down the air-gap at the other end, encounter a considerable reluctance, the whole of the magnetic lines will not go that way; a lot of them will take some shorter cut, although it may be all through air, and you will have some leakage across from limb to limb. I do not say you never have leakage under other circumstances; even with an armature in apparent contact there is always a certain amount of sideway leakage. It depends on the goodness of the contact. And if you widen the air-gaps still further, you will have still more reluctance in the path, and still less magnetism, and still more leakage. Fig. 33 roughly indicates this further stage. The armature will be far less strongly pulled, because, in the first place, the increased reluctance strangles the flow of magnetic lines, so that there are fewer of them in the magnetic cir-

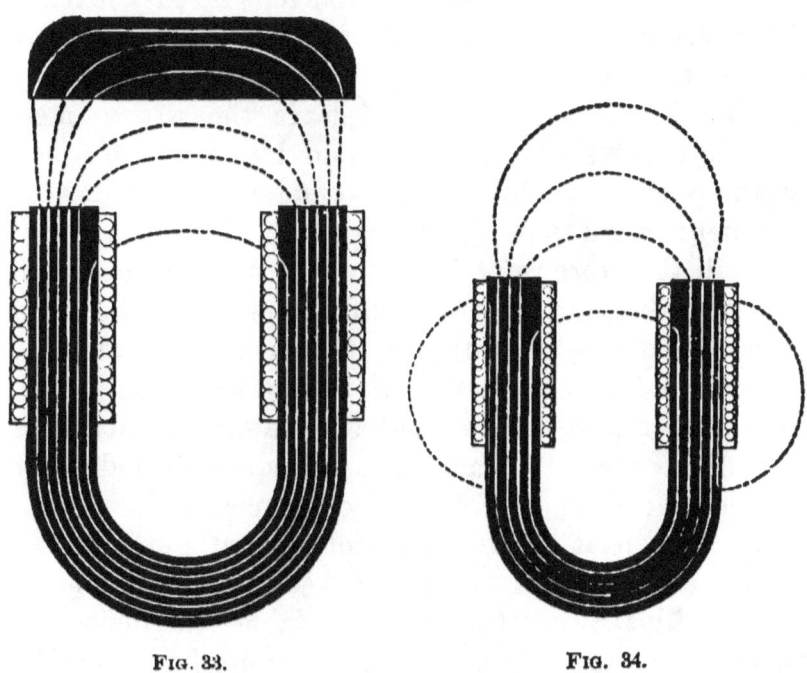

FIG. 33.   FIG. 34.

cuit; and, in the second place, of this lesser number only a fraction reach the armature because of the increased leakage. When you take the armature entirely away the only magnetic lines that go through the iron are those that flow by leakage across the air from the one limb to the other. This is roughly illustrated by Fig. 34, the last of this set.

Leakage across from limb to limb is always a waste of the magnetic lines, so far as useful purposes are concerned. Therefore it is clear that, in order to study the effect of introducing the distance between the armature and the magnet, we have to take into account the leakage; and to calculate the leakage is no easy matter. There are so many considerations that occur as to that which one has to take into account, that it is not easy to choose the right ones and leave the wrong ones. Calculations we must make by and by—they are added as an appendix to this lecture—but for the moment experiment seems to be the best guide.

I will therefore refer, by way of illustrating this question of leakage, to some experiments made by Sturgeon. Sturgeon had a long tubular electromagnet made of a piece of old musket barrel of iron wound with a coil; he put a compass needle about a foot away, and observed the effect. He found the compass needle deflected about 23 degrees; then he got a rod of iron of equal length and put it in at the end, and found that putting it in so that only the end was introduced—in the manner I am now illustrating to you on the table—the deflection increased from 23 degrees to 37 degrees; but when he pushed the iron right home into the gun barrel it went back to nearly 23 degrees. How do you account for that? He had unconsciously increased its facility for leakage when he lengthened out the iron core. And when he pushed the rod right home into the barrel, the extra leakage which was due to the added surface could not and did not occur. There was additional cross-section, but what of that? The additional cross-section is practically of no account. You want to force the magnetism across some 20 inches of air which resists from 300 to 1,000 times as much as iron. What is the use of doubling the section of the iron? You want to reduce the air reluctance, and you have not reduced the air by putting a core into the tube.

There is a paradoxical experiment which we will try next week that illustrates an important principle. If you take a tubular electromagnet and put little pieces of iron into the ends of the iron tube that serves as core, and then magnetize it, the little pieces of iron will try to push themselves out. There is always a tendency to try and increase the completeness of the magnetic circuit; the circuit tends to rearrange itself so as to make it easier for the magnetic lines to go round.

Here is another paradoxical experiment. I have here a bar electromagnet, which we will connect to the wires that bring the exciting current. In front of it, and at a distance from one end of the iron core, is a small compass needle with a feather attached to it as a visible indicator so that when we turn on the current the electromagnet will act on the needle, and you will see the feather turn round. It is acting there at a certain distance. The magnetizing force is mainly spent not to drive magnetism round a circuit of iron, but to force it through the air, flowing from one end of the iron core out into the air, passing by the compass needle, and streaming round again, invisible, into the other end of the iron core. It ought to increase the flow if we can in any way aid the magnetic lines to flow through the air. How can I aid this flow? By putting on something at the other end to help the magnetic lines to get back home. Here is a flat piece of iron. Putting it on here at the hinder end of the core ought to help the flow of magnetic lines. You see that the feather makes a rather larger excursion. Taking away the piece of iron diminishes the effect. So also in experiments on tractive power, it can be proved that the adding of a mass of iron at the far end of a straight electromagnet greatly increases the pulling power at the end that you are working with; while, on the other hand, putting the same piece of iron on the front end as a pole piece greatly diminishes the pull. Here, clamped to the table, is a bar electromagnet excited by the current, and here is a small piece of iron attached to a spring balance by means of which I can measure the pull required to detach it. With the current which I am employing the pull is about two and a half pounds. I now place upon the front end of the core this block of wrought iron; it is itself strongly held on; but the pull which it itself exerts on the small piece of iron is small. Less than half a pound suffices to detach it. I now remove the iron block from the front end of the core and place it upon the hinder end. And now I find that the force required to detach the small piece of iron from the front end is about three and a half pounds instead of two and a half pounds. The front end exerts a bigger pull when there is a mass of iron attached to the hinder end. Why? The whole iron core, including its front end, becomes more highly magnetized, because there is now a better way for the magnetic lines to emerge at the other end and come round to this. In short, we have diminished the magnetic reluctance of the air part of the magnetic circuit, and the flow of magnetic lines in the whole magnetic circuit is thereby improved. So it was also when the mass of iron was placed across the front end of the core; but the magnetic lines streamed away backward from its edges, and few were left in front to act upon the small bit of iron. So the law of magnetic circuit action explains this anomalous behavior. Facts like these have been well known for a long time to those who have studied electromagnets. In Sturgeon's book there is a remark that bar magnets pull better if they are armed with a mass of iron at the distant end, though Sturgeon did not see what we now know to be the explanation of it. The device of fastening a mass of iron to one end of an electromagnet in order to increase the magnetic power of the other end was patented by Siemens in 1862.

We are now in a position to understand the bearing of some curious and important researches made about 40 years ago by Dr. Julius Dub, which, like a great many other good things, lie buried in the back volumes of *Poggendorff's Annalen.* Some account of them is also given in Dr. Dub's now obsolete book, entitled "Elektromagnetismus."

The first of Dub's experiments to which I will refer relates to the difference in behavior between electromagnets with flat and those with pointed pole ends. He formed two cylindrical cores, each six inches long, from the same rod of soft iron, one inch in diameter. Either of these could be slipped into an appropriate magnetizing coil. One of them had the end left flat, the other had its end pointed, or, rather, it was coned down until the flat end was left only half an inch in diameter, possessing therefore only one-fourth of the amount of contact surface which the other core possessed. As an armature there was used another piece of the same soft iron rod, 12 inches long. The pull of the electromagnet on the armature at different distances was carefully measured, with the following results:

Distance apart in inches.	Pull on Flat Pole (lbs.).	Pull on Pointed Pole (lbs.).
0.	3.3	5.2
0.0055	1.1	1.8
0.0110	0.9	0.75
0.0165	0.71	0.50
0.022	0.60	0.42
0.044	0.38	0.20
0.088	0.19	0.09

These results are plotted out in the curves in Fig. 35. It will be seen that in contact, and at very short distances, the reduced pole gave the greater pull. At about ten mils distance there was equality, but at all distances greater than ten mils the flat pole had the advantage. At small distances the concentration of magnetic lines gave, in accordance with the law of traction, the advantage to the reduced pole. But this advantage was, at the greater distances, more than outweighed by the fact that with the greater widths of air-gap the use of the pole with larger face reduced the magnetic reluctance of the gap and promoted a larger flow of magnetic lines into the end of the armature.

Dub's next experiments relate to the employment of polar extensions or pole-pieces attached to the core.

These experiments are so curious, so unexpected, unless you know the reasons why, that I invite your especial attention to them. If an engineer had to make a firm joint between two pieces of metal, and he feared that a mere attachment of one to the other was not adequately strong, his first and most natural impulse would be to

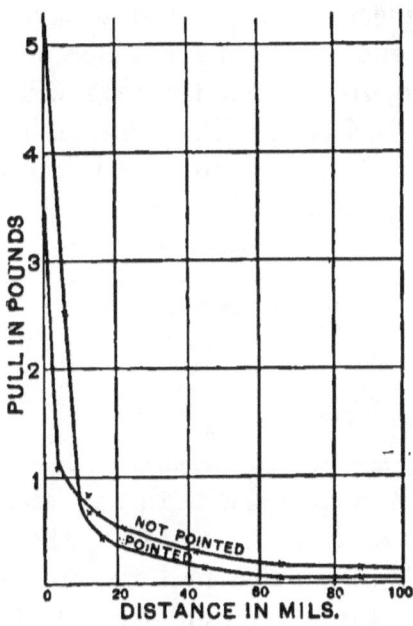

FIG. 35.—CONTRASTED EFFECT OF FLAT AND POINTED POLES.

enlarge the parts that come together—to give one, as it were, a broader footing against the other. And that is precisely what an engineer, if uninstructed in the true principles of magnetism, would do in order to make an electromagnet stick more tightly on to its armature. He would enlarge the ends of one or both. He would add pole-pieces to give the armature a better foothold. Nothing, as you will see, could be more disastrous. Dub employed in these experiments a straight electromagnet having a cylindrical soft iron core, one inch in diameter, twelve inches long; and as armature a piece of the same iron, six inches long. Both were flat ended. Then six pieces of soft iron were prepared of various sizes, to serve as pole-pieces. They could be screwed on at will either to the end of the magnet core or to that of the armature. To distinguish them we will call them by the letters A, B, C, etc. Their dimensions were as follows, the inches being presumably Bavarian inches:

Piece.	Diameter.	Length.
	Inches.	Inches.
A	2	1
B	1¾	1¼
C	1⅝	2
D	2	½
E	1½	1
F	1	2

Of the results obtained with these pieces we will select eight. They are those illustrated by the eight collected sketches in Fig. 36. The pull required to detach was measured, also the attraction exerted at a certain distance apart.

Experiment.	On Magnet.	On Armature.	Traction.	Attraction.
I	None.	None.	48	22
II	D	None.	30	10
III	E	None.	32	11.5
IV	C	None.	35	13.5
V	D	A	20	7.5
VI	None.	B	50	25
VII	None.	D	43	25
VIII	None.	C	50	18

It will be noted that, in every case, putting on a pole-piece to the end of the magnet diminished both the pull in contact and the attraction at a distance; it simply promoted leakage and dissipation of the magnetic lines.

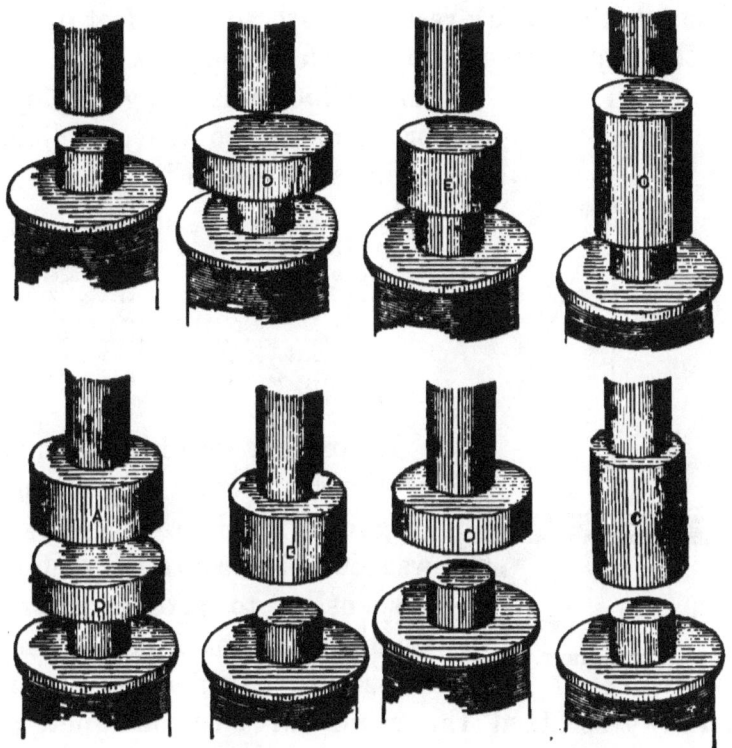

Fig. 36.—Dub's Experiments with Pole-Pieces.

The worst case of all was that in which there were pole-pieces both on the magnet and on the armature. In the last three cases the pull was increased, but here the enlarged piece was attached to the armature, so that it helped those magnetic lines which came up into it to flow back laterally to the bottom end of the electromagnet, while thus reducing the magnetic reluctance of the return path through the air, and so, increasing the total number of magnetic lines, did not spread unduly those that issued up from the end of the core.

The next of Dub's results relate to the effect of adding these pole-pieces to an electromagnet 12 inches long, which was being employed, broadside on, to deflect a distant compass needle (Fig. 37).

Pole-piece used.	Deflection (degrees).
None	34.5
A	42
B	41.5
C	40.5
D	41
E	39
F	38

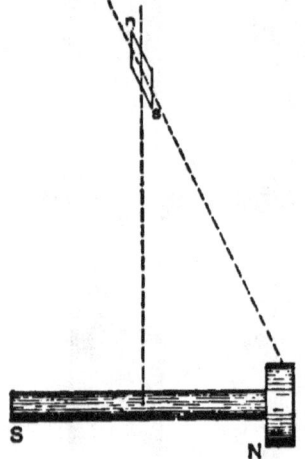

FIG. 37.—DUB'S DEFLECTION EXPERIMENT.

In another set of experiments of the same order a permanent magnet of steel, having poles $n\ s$, was slung horizontally by a bifilar suspension, to give it a strong tendency to set in a particular direction. At a short distance laterally was fixed the same bar electromagnet, and the same pole-pieces were again employed. The results of attaching the pole-pieces at the near end are not very conclusive; they slightly increased the deflection. But in the absence of information as to the distance between the steel magnet and the electromagnet, it is difficult to assign proper values to all the causes at work. The results were:

LECTURES ON THE ELECTROMAGNET. 131

Pole-piece used.	Deflection (degrees).
None.	8.5
A	9.2
B	9.5
C	10
D	8.8

When, however, the pole-pieces were attached to the distant end of the electromagnet, where their effect would undoubtedly be to promote the leakage of mag-

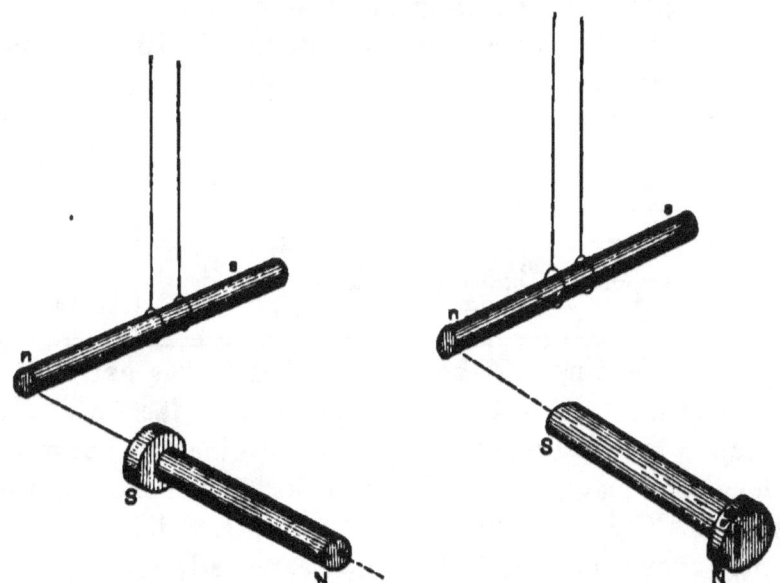

FIG. 38.—DEFLECTING A STEEL MAGNET HAVING BIFILAR SUSPENSION—POLE-PIECE ON NEAR END.

FIG. 39.—DEFLECTING STEEL MAGNET—POLE-PIECE ON DISTANT END.

netic lines into the air at the front end without much affecting the distribution of those lines in the space in front of the pole, the action was more marked.

Pole-piece used.	Deflection (degrees).
None.	8.5
A	10.0
B	10.3
C	10.3
F	10.1

Still confining ourselves to straight electromagnets, I now invite your attention to some experiments made in 1862 by the late Count Du Moncel as to the effect of adding a polar expansion to the iron core. He used as his core a small iron tube, the end of which he could close up with an iron plug, and around which he placed an iron ring which fitted closely on to the pole. He used a special lever arrangement to measure the attraction exercised upon an armature distant in all cases one millimetre from the pole. The results were as follows:

	Without ring on pole.	With ring on pole.
Tubular core alone.............................	11	10
" " with iron plug.................	17	14
Core provided with mass of iron at distant end..	27	25
" " " " with iron plug.......	38	33

After hunting up these researches it was extremely interesting to find that so important a fact had not escaped the observant eye of the original inventor of the electromagnet. In Sturgeon's "Experimental Researches" (p. 113) there is a foot note, written apparently about the year 1832, which runs as follows:

"An electromagnet of the above description, weighing three ounces, and furnished with one coil of wire, supported 14 pounds. The poles were afterward made to expose a large surface by welding to each end of the cylindric bar a square piece of good soft iron; with this alteration only the lifting power was reduced to about five pounds, although the magnet was annealed as much as possible."

We saw that this straight electromagnet, whether used broadside on or end on, could act on the compass needle at some distance from it, and deflect it. In those experiments there was no return path for the magnetic lines that flowed through the iron core save that afforded by the surrounding air. The lines flowed round in wide-sweeping curves from one end to the other, as in Fig. 26; the magnetic field being quite extensive. Now, what will happen if we provide a return path? Suppose I surround the electromagnet with an iron tube of the same length as itself, the lines will flow along in one direction through the core, and will find an easy path back along the outside of the coil. Will the magnet thus jacketed pull more powerfully or less on that little suspended magnet? I should expect it to pull less powerfully, for if the magnetic lines have a good return path here through the iron tube, why should they force themselves in such a quantity to a distance through air in order to get home? No, they will naturally return short back from the end of the core into the tubular iron jacket. That is to say, the action at a distance ought to be diminished by putting on that iron tube outside. Here is the experiment set up. And you see that when I turn on the current my indicating needle is scarcely affected at all. The iron jacket causes that magnet to have much *less* action at a distance. Yet I have known people who actually proposed to use jacketed magnets of this sort in telegraph instruments, and in electric motors, on the ground that they give a bigger pull. You have seen that they produce less action at a distance across air, but there yet remains the question whether they give a bigger pull in contact? Yes, undoubtedly they do; because everything that is helping the magnetism to get round to the other end increases the goodness of the magnetic circuit, and therefore increases the total magnetic flux.

We will try this experiment upon another piece of apparatus, one that has been used for some years at the Finsbury Technical College. It consists of a straight electromagnet set upright in a base-board, over which is erected a light gallows of wood. Across the frame of the gallows goes a winch, on the axle of which is a small pulley with a cord knotted to it. To the lower end of the cord is hung a common spring balance, from the hook of which depends a small horizontal disc of iron to act as an armature. By means of the winch I lower this disc down to the top of the electromagnet. The current is turned on: the disc is attracted. On winding up the winch I increase the upper pull until the disc is detached. See, it required about nine pounds to pull it off. I now slip over the electromagnet, without in any way attaching it, this loose jacket of iron—a tube, the upper end of which stands flush with the upper polar surface. Once more I lower the disc, and this time it attaches itself at its middle to the central pole, and at its edges to the tube. What force will now be required to detach it? The tube weighs about one-half pound, and it is not fixed at the bottom. Will $9\frac{1}{2}$ pounds suffice to lift the disc? By no means. My balance only measures up to 24 pounds, and even that pull will not suffice to detach the disc. I know of one case where the pull of the straight core was increased 16-fold by the mere addition of a good return path of iron to complete the magnetic circuit. It is curious how often the use of a tubular jacket to an electromagnet has been reinvented. It dates back to about 1850 and has been variously claimed for Romershausen, for Guillemin, and for Fabre. It is described in Davis' "Magnetism," published in Boston in 1855. About sixteen years ago Mr. Faulkner, of Manchester, revived it under the name of the Altandae electromagnet. A discussion upon jacketed electromagnets took place in 1876 at the Society of Telegraph Engineers; and in the same year Professor Graham Bell used the same form of electromagnet in the receiver of the telephone which he exhibited at the Centennial Exhibition. But the jacketed form is not good for anything except increasing the tractive power. Jacketing an electromagnet which already possesses a return circuit of iron is an absurdity. For this reason the proposal made by one inventor to put iron tubes outside the coils of a horseshoe electromagnet is one to be avoided.

We will take another paradox, which equally can be explained by the principle of the magnetic circuit. Suppose you take an iron tube as an interior core; suppose you cut a little piece off the end of it; a mere ring of the same size. Take that little piece and lay it down on the end. It will be struck with a certain amount of pull. It will pull off easily. Take that same round piece of iron, put it on edgewise, where it only touches one point of the circumference, and it will stick on a good deal tighter, because it is there in a position to increase the magnetic flow of the magnetic lines. Concentrating the flow of magnetic lines over a small surface of contact increases **B** at that point and **B**^{2}, integrated over the lesser area of the contact, gives a total bigger pull than is the case when the edge is touched all round against the edge of the tube.

Here is a still more curious experiment. I use a cylindrical electromagnet set up on end, the core of which has at the top a flat, circular polar surface, about two inches in diameter. I now take a round disc of thin iron—ferrotype or tin-plate will answer quite well—which is a little smaller than the polar face. What will happen when this disc is laid down flat and centrally on the polar face? Of course you will say that it will stick tightly on. If it does so, the magnetic lines which come in through its under surface will pass through it and come out on its upper surface in large quantities. It is clear that they cannot all, or even any considerable proportion of them, emerge sidewise through the edges of the thin disc, for there is not substance enough in the disc to carry so many magnetic lines. As a matter of fact the magnetic lines do come through the disc and emerge on its upper surface, making indeed a magnetic field over its upper surface that is nearly as intense as the magnetic field beneath its under surface. If the two magnetic fields were exactly of equal strength, the disc ought not to be attracted either way. Well, what is the fact? The fact, as you see now that the current has been turned on, is that the disc absolutely refuses to lie down on the top of the pole. If I hold it down with my finger, it actually

FIG. 40.—EXPERIMENT WITH TUBULAR CORE AND IRON RING.

bends itself up and requires force to keep it down. I lift my finger, and over it flies. It will go anywhere in its effort to better the magnetic circuit rather than lie flat on top of the pole.

Next I invite your attention to some experiments, originally due to Von Kolke, published in the *Annalen* 40 years ago, respecting the distribution of the magnetic lines where they emerge from the polar surface of an electromagnet. I cannot enumerate them all, but will merely illustrate them by a single example. Here is a straight electromagnet with a cylindrical, flat-ended core (Fig. 41). In what way will the magnetic lines be distributed over it at the end? Fig. 26 illustrates roughly the way in which, when there is no return path of iron, the magnetic lines leak through the air. The main leakage is

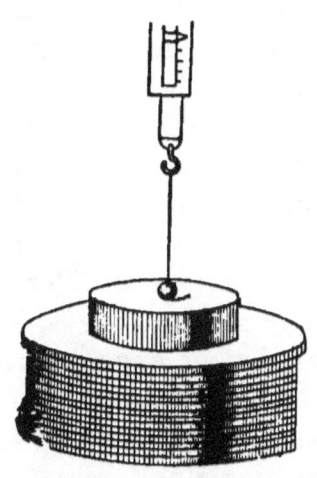

FIG. 41—EXPLORING POLAR DISTRIBUTION WITH SMALL IRON BALL.

through the ends, though there is some at the sides also. Now the question of the end distribution we shall try by using a small bullet of iron, which will be placed at different points from the middle to the edge, a spring balance being employed to measure the force required to detach it. The pull at the edge is much stronger than at the middle, at least four or five times as great. There is a regular increase of pull from the middle to the edge. The magnetic lines, in trying to complete their own circuit, flow most numerously in that direction where they can go furthest through iron on their journey. They leak out more strongly at all edges and corners of a polar surface. They do not flow out so strongly at the middle of the end surface, otherwise they would have to go through a larger air circuit to get back home. The iron is consequently more saturated round the edge than at the middle; therefore, with a very small magnetizing force, there is a great disproportion between pull at the middle and that at the edges. With a very large magnetizing force you do not get the same disproportion, because if the edge is already far saturated you cannot by applying higher magnetizing power increase its magnetization much, but you can still force more lines through the middle. The consequence is, if you plot out the results of a succession of experiments of the pull at different points, the curves obtained are, with larger magnetizing forces, more nearly straight than are those obtained with small magnetizing forces. I have known cases where the pull at the edge was six or seven times as great as in the middle with a small magnetizing power, but with larger power not more than two or three times as great, although, of course, the pull all over was greater You can easily observe this distinction by merely putting a polished iron ball upon the end of the electromagnet, as in Fig. 42. The ball at once rolls to the edge and will not stay at the middle. If I take a larger two-pole electromagnet (like Fig. 11), what will the case now be? Clearly the shortest path of the magnetic lines through the air is the path just across from the edge of one polar surface to the edge of the other between the poles. The lines are most dense in the region where they arch over in as short an arch as possible, and they will be less dense along the longer paths, which arch more widely over. Therefore, as there is a greater tendency to leak from the inner edge of one pole to the inner edge of the other, and less tendency to leak from the outer edge of one to the outer edge of the other, the biggest pull ought to be on the inner edges of the pole. We will now try it.

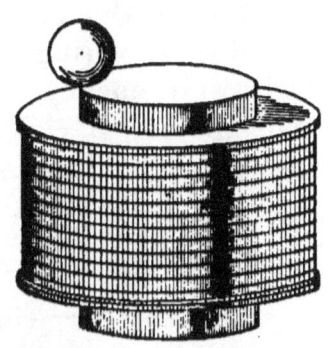

Fig. 42.—Iron Ball Attracted to Edge of Polar Face.

On putting the iron ball anywhere on the pole it immediately rolls until it stands perpendicularly over the inner edge.

The magnetic behavior of little iron balls is very curious. A small round piece of iron does not tend to move at all in the most powerful magnetic field if that magnetic field is uniform. All that a small ball of iron tends to do is to move from a place where the magnetic field is weak to a place where the magnetic field is strong. Upon that fact depends the construction of several important instruments, and also certain pieces of electromagnetic mechanism.

In order to study this question of leakage, and the relation of leakage to pull, still more incisively, I devised some time ago a small experiment with which a group of my students at the Technical College have been diligently experimenting. Here (Fig. 43) is a horseshoe electromagnet. The core is of soft wrought iron, wound with a known number of turns of wire. It is provided with an armature. We have also wound on three little exploring coils, each consisting of five turns of wire only, one, *C*, right down at the bottom on the bend; another, *B*, right round the pole, close up to the armature, and a third, *A*, around the middle of the armature. The object of these is to ascertain how much of the magnetism which was created in the core by the magnetizing power of these coils ever got into the armature. If the armature is at a considerable distance away, there is naturally a great deal of leakage. The coil *C* around the bend at the bottom is to catch all the magnetic lines that go through the iron; the coil *B* at the poles is to catch all that have not leaked outside before the magnetism has crossed the joint; while the coil *A*, right around the middle of the armature, catches all the lines that actually pass into the armature, and pull at it. We measure by means of the ballistic galvanometer and these three exploring coils how much magnetism gets into the armature at different distances, and are able thus to determine the leakage, and compare these amounts with the calculations made, and with the

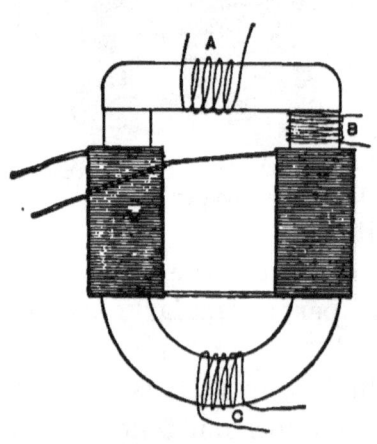

FIG. 43.—EXPERIMENT ON LEAKAGE OF ELECTROMAGNET.

attractions at different distances. The amount of magnetism that gets into the armature does not go by a law of inverse squares, I can assure you, but by quite other laws. It goes by laws which can only be expressed as particular cases of the law of the magnetic circuit. The most important element of the calculations, indeed, in many cases is the amount of percentage of leakage that must be allowed for. Of the magnitude of this matter you will get a very good idea by the result of these experiments following.

The iron core is 13 millimetres in diameter, and the coil consists of 178 turns. The first swing of the galvanometer when the current was suddenly turned on or off measured the number of magnetic lines thereby sent through, or withdrawn from, the exploring coil that is at the time joined to the galvanometer. The currents used varied from 0.7 of an ampère to 5.7 ampères. Six sets of experiments were made, with the armature at different distances. The numerical results are given below:

I.—WITH WEAK CURRENT (0.7 AMPÈRE).

	A	B	C
In contact	12,506	13,870	14,190
Armature distance 1 mm	1,552	2,163	3,786
Armature distance 2 mm	1,149	1,487	2,839
Armature distance 5 mm	1,014	1,081	2,028
Armature distance 10 mm	670	1,014	1,690
Removed	—	675	1,352

II.—STRONGER CURRENT (1.7 AMPÈRES).

	A	B	C
In contact..	18,240	19,590	20,283
Armature distance. 1 mm	2,570	3,381	5,408
2 mm	2,366	2,839	5,073
5 mm	1,352	2,299	5,949
10 mm	811	1,352	3,381
Removed	—	1,308	3,041

III.—STILL STRONGER CURRENT (3.7 AMPÈRES).

	A	B	C
In contact	20,940	22,280	22,960
Armature distance. 1 mm	5,610	7,568	11,831
2 mm	4,597	6,722	9,802
5 mm	2,569	3,245	7,436
10 mm	1,149	2,704	7,098
Removed	—	2,366	6,427

IV.—STRONGEST CURRENT (5.7 AMPÈRES).

	A	B	C
In contact	21,980	23,660	24,040
Armature distance. 1 mm	8,110	10,810	17,220
2 mm	5,611	8,464	15,886
5 mm	4,056	5,273	12,627
10 mm	2,029	4,057	10,142
Removed	—	3,581	9,795

These numbers may be looked upon as a kind of numerical statement of the facts roughly depicted in Figs. 31 to 34. The numbers themselves, so far as they relate to the measurements made (1) in contact, (2) with gaps of one millimetre breadth, are plotted out on Fig. 44, there being three curves, A, B and C, for the measurements made when the armature was in contact, and three others, A_1, B_1, C_1, made at the one millimetre distance. A dotted line gives the plotting of the numbers for the coil C, with different currents, when the armature was removed.

On examining the numbers in detail we observe that the largest number of magnetic lines forced round the bend of the iron core, through the coil C, was 24,040 (the cross-section being a little over one square centi-

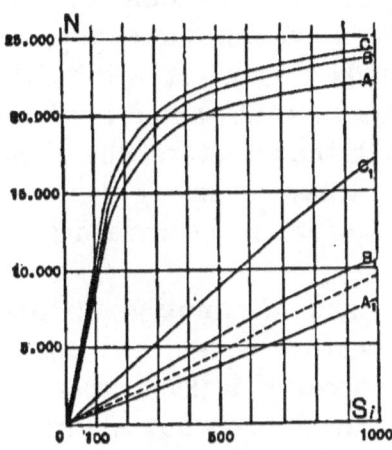

FIG. 44.—CURVES OF MAGNETIZATION PLOTTED FROM PRECEDING.

metre), which was when the armature was in contact. When the armature was away the same magnetizing power only evoked 9,795 lines. Further, of those 24,040, 23,660 (or $98\frac{1}{2}$ per cent.) came up through the polar surfaces of contact, and of those again 21,980 (or $92\frac{1}{2}$ per cent. of the whole number) passed through the armature. There was leakage, then, even when the armature was in contact, but it amounted to only $7\frac{1}{2}$ per cent. Now, when the armature was moved but one millimetre (*i. e.*, one twenty-fifth of an inch) away, the presence of the air-gaps had this great effect, that the total magnetic flux was at once choked down from 24,040 to 17,220. Of that number only 10,810 (or 61 per cent.) reached the polar surfaces, and only 8,110 (or 47 per cent. of the total number) succeeded in going through the armature. The leakage in this case was 53 per cent.! With a two millimetre gap the leakage was 65 per cent. when the strongest current was used. It was 68 per cent. with a five millimetre gap, and 80 per cent. with a 10 millimetre gap. It will further be noticed that while a current of 0.7 ampère sufficed to send 12,506 lines through the armature when it was in contact, a current eight times as strong could only succeed in sending 8,110 lines when the armature was distant by a single millimetre.

Such an enormous diminution in the magnetic flux through the armature, consequent upon the increased reluctance and increased leakage occasioned by the presence of the air-gaps, proves how great is the reluctance offered by air, and how essential it is to have some practical rules for calculating reluctances and estimating leakages to guide us in designing electromagnets to do any given duty.

The calculation of magnetic reluctances of definite portions of a given material are now comparatively easy, and, thanks to the formulæ of Prof. Forbes, it is now possible in certain cases to estimate leakages. Of these methods of calculation an abstract will be given in the appendix to this lecture. I have, however, found Forbes' rules, which were intended to aid the design of dynamo machines, not very convenient for the common cases of electromagnets, and have therefore cast about to discover some more apposite mode of calculation. To predetermine the probable percentage of leakage one must first distinguish between those magnetic lines which go usefully through the armature (and help to pull it) and those which go astray through the surrounding air and are wasted so far as any pull is concerned. Having set up this distinction, one then needs to know the relative magnetic conductance, or *permeance*, along the path of the useful lines, and that along the innumerable paths of the wasted lines of the stray field. For (as every electrician accustomed to the problems of shunt circuits will recognize) the quantity of lines that go respectively along the useful and wasteful paths will be directly proportional to the conductances (or permeances) along those paths, or will be inversely proportional to the respective resistances along those paths. It is customary in electromagnetic calculations to employ a certain coefficient of allowance for leakage, the symbol for which is v, such that when we know the number of magnetic lines that are wanted to go through the armature we must allow for v times as many in the magnet core. Now, if u represents permeance along the useful path, and w the permeance of all the waste paths along the stray field, the total flux will be to the useful flux as $u + w$ is to u. Hence the coefficient of allowance for leakage v is equal to $u + w$ divided by u. The only real difficulty is to calculate u and w. In general u is easily calculated; it is the reciprocal of the sum of all the magnetic reluctances along the useful path from pole to pole. In the case of the electromagnet used in the experiments last described, the magnetic reluctances along the useful path are three in number, that of the iron of the armature and those of the two air-gaps. The following formula is applicable,

$$\text{reluctance} = \frac{l_1}{A_1 \mu_1} + \frac{2l_2}{A_2},$$

if the data are specified in centimetre measure, the suffixes 1 and 2 relating respectively to the iron and to the air. If the data are specified in inch measures the formula becomes

$$\text{reluctance} = 0.3132 \left\{ \frac{l''_1}{A''_1 \mu_1} + \frac{2l''_2}{A''_2} \right\}$$

But it is not so easy to calculate the reluctance (or its reciprocal, the permeance) for the waste lines of the stray field, because the paths of the magnetic lines spread out so extraordinarily and bend round in curves from pole to pole.

Fig. 45 gives a very fair representation of the spreading of the lines of the stray field that leaks across between the two limbs of a horseshoe electromagnet made of round iron. And for square iron the flow is much the same, except that it is concentrated a little by the corners of the metal. Forbes' rules do not help us here. We want a new mode of considering the subject.

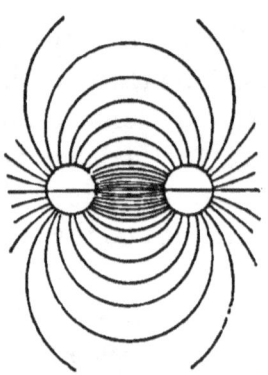

FIG. 45.—CURVES OF FLOW OF MAGNETIC LINES IN AIR FROM ONE CYLINDRICAL POLE TO ANOTHER.

The problems of flow, whether of heat, electricity or of magnetism, in space of three dimensions, are not among the most easy of geometrical exercises. However, some of them have been worked out, and may be made applicable to our present need. Consider, for example, the electrical problem of finding the resistance which an indefinitely extended liquid (say a solution of sulphate of copper of given density) offers when acting as a conductor of electric currents flowing across between two indefinitely long parallel cylinders of copper. Fig. 45 may be regarded as representing a transverse section of such an arrangement, the sweeping curves representing lines of flow of current. In a simple case like this it is possible to find an accurate expression for the resistance (or of the conductance) of a layer or stratum of unit thickness. It depends on the diameters of the cylinders, on their distance apart, and on the specific conductivity of the medium. It is not by any means proportional to the distance between them, being, in fact, almost independent of the distance, if that is greater than 20 times the perimeter of either cylinder. Neither is it even approximately proportional to the perimeter of the cylinders except in those cases when the shortest distance between them is less than a tenth part of the perimeter of either. The resistance, for unit length of the cylinders, is, in fact, calculated out by the rather complex formula:

$$R = \frac{1}{\pi/\mu} \log. \text{nat. } h;$$

Where
$$h = \frac{2a}{b + 2a - \sqrt{b^2 + 4ab}}$$

the symbol a standing for the radius of the cylinder; b for the shortest distance separating them; μ for the permeability, or in the electric case the specific conductivity of the medium.

Now, I happened to notice, as a matter that greatly simplifies the calculation, that if we confine our attention to a transverse layer of the medium of given thickness, the resistance between the two bits of the cylinders in that layer depends on the ratio of the shortest distance separating them to their periphery, and is independent of the absolute size of the system. If you have the two cylinders an inch round and an inch between them, then the resistance of the slab of medium (of given thickness) in which they lie will be the same as if they were a foot round and a foot apart. Now that simplifies matters very much, and thanks to my friend and former chief assistant, Dr. R. Mullineux Walmsley, who devoted himself to this troublesome calculation, I am able to give you, in tabular form, the magnetic resistances within the limits of proportion that are likely to occur.

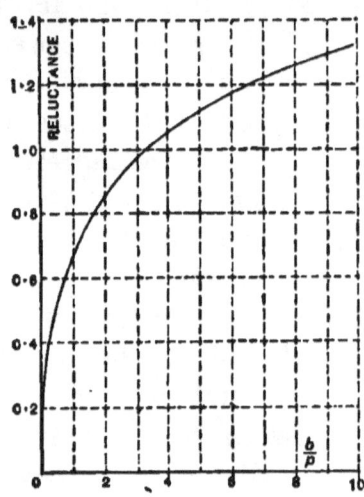

Fig. 46.—Diagram of Leakage Reluctances.

TABLE VIII.—MAGNETIC RELUCTANCE OF AIR BETWEEN TWO PARALLEL CYLINDRICAL LIMBS OF IRON.

$\dfrac{b}{p}$ Ratio of least distance apart to perimeter.	Magnetic reluctance in C. G. S. units = the magneto-motive force ÷ total magnetic flux.		Magnetic reluctance in inch units = the ampère turns ÷ the total magnetic flux. Slab = 1 inch thick.	
	Reluctance.	Permeance.	Reluctance.	Permeance.
0.1	0.2461	4.063	0.0771	12.968
0.2	0.3404	2.938	0.1066	9.377
0.3	0.4084	2.449	0.1280	7.815
0.4	0.4628	2.161	0.1450	6.897
0.5	0.5084	1.967	0.1593	6.278
0.6	0.5479	1.825	0.1717	5.825
0.8	0.6140	1.629	0.1924	5.198
1.0	0.6681	1.497	0.2093	4.777
1.2	0.7144	1.400	0.2238	4.571
1.4	0.7550	1.324	0.2365	4.228
1.6	0.7908	1.265	0.2476	4.039
1.8	0.8220	1.217	0.2575	3.883
2.0	0.8511	1.202	0.2667	3.750
4.0	1.0500	0.952	0.3290	3.040
6.0	1.1710	0.854	0.3669	2.726
8.0	1.2624	0.792	0.3955	2.528
10.0	1.3250	0.755	0.4151	2.409

NOTE.—In the above table, unit length of cylinders is assumed (1 centimetre in columns 2 and 3; 1 inch in columns 4 and 5); the flow of magnetic lines being reckoned as in a slab of infinite extent and of unit thickness. Symbols: p = perimeter of cylinder; b = shortest distance between cylinders. In columns 2 and 3 the unit reluctance is that of a centimetre cube of air. In columns 4 and 5 the unit reluctance is so chosen (as in the rest of these lectures wherever such measures are used) that the reduction of ampère turns to magneto-motive force by multiplying by $4\pi \div 10$ is avoided. This will make the reluctance of the inch cube of air equal to $10 \div 4\pi \div 2.54 = 0.3132$, and its permeance as 3.1931.

The numbers from columns 1 and 2 of the preceding table are plotted out graphically in Fig. 46 for more convenient reference. As an example of the use of the table we will take the following:

EXAMPLE.—Find the magnetic reluctance and permeance between two parallel iron cores of one inch diameter and nine inches long, the least distance between them being 2⅜ inches. Here $b = 2.375$; $p = 3.1416$; $b \div p = 0.756$. Reference to the table shows (by interpolation) that the reluctance and permeance for unit thickness of slab are respectively 0.183 and 5.336. For nine inches thickness they will therefore be 0.021 and 48.02 respectively.

When the permeance across between the two limbs is thus approximately calculable, the waste flux across the space is estimated by multiplying the permeance so found by the *average* value of the difference of magnetic potential between the two limbs. And this, if the yoke which unites the limbs at their lower end is of good solid iron, and if the parallel cores offer little magnetic reluctance as compared with the reluctance of the useful paths, or of that of the stray field, may be simply taken as half the ampère turns (or, if centimetre measures are used, multiply by 1.2566).

The method here employed in estimating the reluctance of the waste field is of course only an approximation; for it assumes that the leakage takes place only in the planes of the slabs considered. As a matter of fact there is always some leakage out of the planes of the slabs. The real reluctance is always therefore somewhat less, and the real permeance somewhat greater, than that calculated from Table VIII.

For the electromagnets used in ordinary telegraph instruments the ratio of b to p is not usually very different from unity, so that for them the permeance across from limb to limb per inch length of core is not very far from 5.0, or nearly twice the permeance of an inch cube of air.

We are now in a position to see the reason for a curious statement of Count Du Moncel which for long puzzled me. He states that he found, using distance apart of one millimetre, that the attraction of a two-pole electromagnet for its armature was less when the armature was presented laterally than when it was placed in front of the pole-ends, in the ratio of 19 to 31. He does not specify in the passage referred to what was the shape of either the armature or the cores. If we assume that he was referring to an electromagnet with cores of the usual sort—round iron with flat ends, presumably like Fig. 11—then it is evident that the air-gaps, when the armature is presented sidewise to the magnet, are really greater than when the armature is presented in the usual way, owing to the cylindric curvature of the core. So, if at equal measured distance the reluctance in the circuit is greater, the magnetic flux will be less and the pull less.

It ought also now to be evident why an armature made of iron of a flat rectangular section, though when in contact it sticks on tighter edgewise, is at a distance attracted more powerfully if presented flatwise. The gaps, when it is presented flatwise (at an equal least distance apart), offer a lesser magnetic reluctance.

Another obscure point also becomes explainable, namely, the observation by Lenz, Barlow, and others, that the greatest amount of magnetism which could be imparted to long iron bars by a given circulation of electric current was (nearly) proportional, not to the cross-sectional area of the iron, but to its surface! The explanation is this: Their magnetic circuit was a bad one, consisting of a straight rod of iron and of a return path through air. Their magnetizing force was being in reality expended not so much on driving magnetic lines through iron (which is readily permeable), but on driving the magnetic lines through air (which is, as we know, much less permeable), and the reluctance of the return paths through the air is—when the distance from one to the other of the exposed end parts of the bar is great compared with its periphery—very nearly proportional to that periphery, that is to say, to the exposed surface.

Another opinion on the same topic was that of Prof. Müller, who laid down the law that for iron bars of equal length, and excited by the same magnetizing power, the amount of magnetism was proportional to the square root of the periphery. A vast amount of industrious scientific effort has been expended by Dub, Hankel, Von Feilitzsch, and others on the attempt to verify this "law." Not one of these experimenters seems to have had the faintest suspicion that the real thing which determined the amount of magnetic flow was not the iron, but the reluctance of the return path through air. Von Feilitzsch plotted out the accompanying curves (Fig. 47), from which he drew the inference that the law of the square root of the periphery was established. The very straightness of these curves shows that in no case had the iron become so much

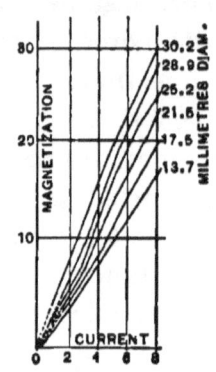

FIG. 47. — VON FEILITZSCH'S CURVES OF MAGNETIZATION OF RODS OF VARIOUS DIAMETERS.

magnetized as to show the bend that indicates approaching saturation. Air, not iron, was offering the main part of the resistance to magnetization in the whole of these experiments. I draw from the very same curves the conclusion that the magnetization is not proportional to the square root of the periphery, but is more nearly proportional to the periphery itself; indeed, the angles at which the different curves belonging to the different peripheries rise show that the amount of magnetism is very nearly as the surface. Observe here we are not dealing with a closed magnetic circuit where section comes into account; we are dealing with a bar in which the magnetism can only get from one end to the other by leaking all round into the air. If, therefore, the reluctance of the air path from one end of the bar to the other is proportional to the surface, we should get some curves very like these; and that is exactly what happens. If you have a solid, of a certain given geometrical form, standing out in the middle of space, the conductance which the space around it (or rather the medium filling that space) offers to the magnetic lines flowing through it, is practically proportional to the surface. It is distinctly so for similar geometrical solids, when they are relatively small as compared with the distance between them. Electricians know that the resistance of the liquid between two small spheres, or two small discs of copper immersed in a large bath of sulphate of copper, is practically independent of the distance between them, provided they are not within ten diameters, or so, of one another. In the case of a long bar we may treat the distance between the protruding ends as sufficiently great to make an approximation to this law hold good. Von Feilitzsch's bars were, however, not so long that the average value of the length of path from one end surface to the other end surface, along the magnetic lines, was infinitely great as compared with the periphery. Hence the departure from exact proportionality to the surface. His bars were 9.1 centimetres long, and the peripheries of the six were respectively 94.9, 90.7, 79.2, 67.6, 54.9 and 42.9 millimetres.

It has long been a favorite idea with telegraph engineers that a long-legged electromagnet in some way possessed a greater "projective" power than a short-legged one; that, in brief, a long-legged magnet could attract an armature at a greater distance from its poles than could a short-legged one made with iron cores of the same section. The reason is not far to seek. To project or drive the magnetic lines across a wide intervening air-gap requires a large magnetizing force on account of the great reluctance, and the great leakage in such cases. And the great magnetizing force cannot be got with short cores, because there is not, with short cores, a sufficient length of iron to receive all the turns of wire that are in such a case essential. The long leg is wanted simply to carry the wire necessary to provide the requisite circulation of current.

We now see how, in designing electromagnets, the length of the iron core is really determined; it must be long enough to allow of the winding upon it of the wire which, without overheating, will carry the ampère turns of exciting current which will suffice to force the requisite number of magnetic lines (allowing for leakage) across the reluctances in the useful path. We shall come back to this matter after we have settled the mode of calculating the quantity of wire that is required.

Being now in a position to calculate the additional magnetizing power required for forcing magnetic lines across an air-gap, we are prepared to discuss a matter that has been so far neglected, namely, the effect on the reluctance of the magnetic circuit of joints in the iron. Horseshoe electromagnets are not always made of one piece of iron bent round. They are often made, like Fig. 11, of two straight cores shouldered and screwed, or riveted into a yoke. It is a matter purely for experiment to determine how far a transverse plane of section across the iron obstructs the flow of magnetic lines. Armatures, when in contact with the cores, are never in perfect contact, otherwise they would cohere without the application of any magnetizing force; they are only in imperfect contact, and the joint offers a considerable magnetic reluctance.

This matter has been examined by Prof. J. J. Thomson and Mr. Newall, in the Cambridge Philosophical Society's *Proceedings*, in 1887; and recently more fully by Prof. Ewing, whose researches are published in the *Philosophical Magazine* for September, 1888. Ewing not only tried the effect of cutting and of facing up with true plane surfaces, but used different magnetizing forces, and also applied various external pressures to the joint. For our present purpose we need not enter into the questions of external pressures, but will summarize the results which Ewing found when his bar of wrought iron was cut across by section planes, first into two pieces, then into four, then into eight. The apparent permeability of the bar was reduced at every cut.

TABLE IX.—EFFECT OF JOINTS IN WROUGHT IRON BAR (NOT COMPRESSED).

H	B				Mean thickness of equivalent air-space for one cut.		Thickness of iron of equivalent reluctance per cut.	
	Solid.	Cut in two.	In four.	In eight.	Centi-metres.	Inches.	Centi-metres.	Inches.
7.5	8,500	6,900	4,809	2,600	0.0036	0.0014	4.	1.57
15	13,400	11,550	8,900	5,550	0.0030	0.0012	2.53	1.00
30	15,350	14,550	12,940	9,800	0.0020	0.0008	1.10	0.433
50	16,400	15,950	15,000	13,300	0.0013	0.0005	0.43	0.169
70	17,100	16,840	16,120	15,200	0.0009	0.0004	0.22	0.087

Suppose we are working with the magnetization of our iron pushed to about 16,000 lines to the square centimetre (*i. e.*, about 150 pounds per square inch, traction), requiring a magnetizing force of about H = 50; then, referring to the table, we see that each joint across the iron offers as much reluctance as would an air-gap 0.0005 of an inch in thickness, or adds as much reluctance as if an additional layer of iron about one-sixth of an inch thick had been added. With small magnetizing forces the effect of having a cut across the iron with a good surface on it is about the same as though you had introduced a layer of air one six-hundredth of an inch thick, or as though you had added to the iron circuit about one inch of extra length. With large magnetizing forces, however, this disappears, probably because of the attraction of the two surfaces across that cut. The stress in the magnetic circuit with high magnetic forces running up to 15,000 or 20,000 lines to the square centimetre will of itself put on a pressure of 130 to 230 pounds to the square inch, and so these resistances are considerably reduced; they come down in fact to about one-twentieth of their initial value. When Ewing specially applied compressing forces, which were as large as 670 pounds to the square inch, which would of themselves ordinarily, in a continuous piece of iron, have diminished the magnetizability, he found the diminution of the magnetizability of iron itself was nearly compensated for by the better conduction of the cut surface. The old surface, cut and compressed in that way, closes up as it were, magnetically — does not act like a cut at all; but at the same time you lose just as much as you gain, because the iron itself becomes less magnetizable.

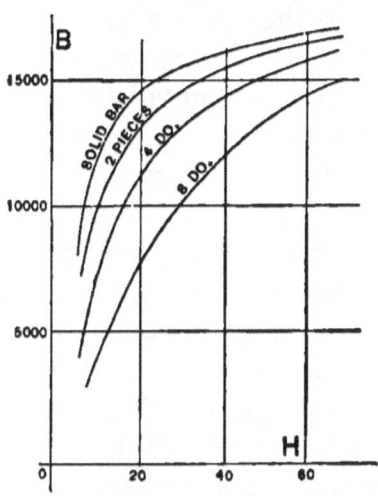

FIG. 48.—EWING'S CURVES FOR EFFECT OF JOINTS.

The above results of Ewing's are further represented by the curves of magnetization drawn in Fig. 48. When the faces of a cut were carefully surfaced up to true planes, the disadvantageous effect of the cut was reduced considerably, and, under the application of a heavy external pressure, almost vanished.

I have several times referred to experimental results obtained in past years, principally by German and

French workers, buried in obscurity in the pages of foreign scientific journals. Too often, indeed, the scattered papers of the German physicists are rendered worthless or unintelligible by reason of the omission of some of the data of the experiments. They give no measurements perhaps of their currents, or they used an uncalibrated galvanometer, or they do not say how many windings they were using in their coils; or perhaps they give their results in some obsolete phraseology. They are extremely addicted to informing you about the "magnetic moments" of their magnets. Now the magnetic moment of an electromagnet is the one thing that one never wants to know. Indeed the magnetic moment of a magnet of any kind is a useless piece of information, except in the case of bar magnets of hard steel that are to be used in the determination of the horizontal component of the earth's magnetic force. What one does want to know about an electromagnet is the number of magnetic lines flowing through its circuit, and this the older researches rarely afford the means of ascertaining. Nevertheless, there are some investigations worthy of study to which time will now only permit me very briefly to allude. These are the researches of Dub on the effect of thickness of armatures, and those of Nicklès and of Du Moncel on the lengths of armatures. Also those of Nicklès on the effect of width between the two limbs of the horseshoe electromagnet.

I can only now describe some experiments of Von Feilitzsch upon the vexed question of tubular cores, a matter touched by Sturgeon, Pfaff, Joule, Nicklès, and later by Du Moncel. To examine the question whether the inner part of the iron really helps to carry the magnetism, Von Feilitzsch prepared a set of thin iron tubes which could slide inside one another. They were all 11 centimetres long, and their peripheries varied from 6.12 centimetres to 9.7 centimetres. They could be pushed within a magnetizing spiral to which either small or large currents could be applied, and their effect in deflecting a magnetic needle was noted, and balanced by means of a compensating steel magnet, from the position of which the forces were reckoned and the magnetic moments calculated out. As the tubes were of equal lengths, the magnetization is approximately proportional to the magnetic moment. The outermost tube was first placed in the spiral, and a set of observations made; then the tube of next smaller size was slipped into it and another set of observations made; then a third tube was slipped in until the whole of the seven were in use. Owing to the presence of the outer tube in all the experiments, the reluctance of the air return paths was alike in every case. The curves given in Fig. 49 indicate the results.

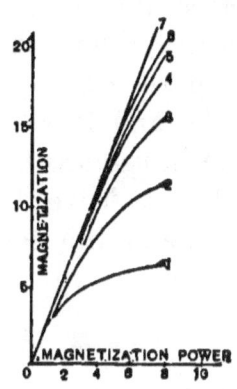

FIG. 49.—VON FEILITZSCH'S CURVES OF MAGNETIZATION OF TUBES.

The lowest curve is that corresponding to the use of the first tube alone. Its form, bending over and becoming nearly horizontal, indicates that with large magnetizing power it became nearly saturated. The second curve corresponds to the use of the first tube with the second within it. With greater section of iron saturation sets in at a later stage. Each successive tube adds to the capacity for carrying magnetic lines, the beginning of saturation being scarcely perceptible, even with the highest magnetizing power, when all seven tubes were used. All the curves have the same initial slope. This indicates that with small magnetizing forces, and when even the least quantity of iron was present, when the iron was far from saturation, the main resistance to magnetization was that of the air paths, and it was the same whether the total section of iron in use was large or small.

I must leave till my next lecture the rules relating to the determination of the windings of copper wire on the cores.

APPENDIX TO LECTURE II.

CALCULATION OF EXCITATION, LEAKAGE, ETC.

Symbols used.

- N = the whole number of magnetic lines (C.G.S. definition of magnetic lines, being one line per square centimetre to represent intensity of a magnetic field, such that there is one dyne on unit magnetic pole) that pass through the magnetic circuit. Also called the *magnetic flux*.
- B = the number of magnetic lines per square centimetre in the iron; also called the *induction*, or the internal magnetization.
- $B_{\prime\prime}$ = the number of magnetic lines per square inch in the iron.
- H = the magnetic force or intensity of the magnetic field, in terms of the number of magnetic lines to the square centimetre that there would be in air.
- $H_{\prime\prime}$ = the magnetic force, in terms of the number of magnetic lines that there would be to the square inch, in air.
- μ = the *permeability* of the iron, etc.; that is its magnetic conductivity or multiplying power for magnetic lines.
- A = area of cross-section, in square centimetres.

A'' = area of cross-section, in square inches.
l = length, in centimetres.
l'' = length, in inches.
S = number of spirals or turns in the magnetizing coil.
i = electric current, expressed in ampères.
v = coefficient of allowance for leakage; being the ratio of the whole magnetic flux to that part of it which is usefully applied. (It is always greater than unity.)

Relations of units.

1 inch = 2.54 centimetres;
1 centimetre = 0.3937 inch.
1 square inch = 6.45 square centimetres;
1 square centimetre = 0.1550 square inch.
1 cubic inch = 16.39 cubic centimetres;
1 cubic centimetre = 0.0610 cubic inch.

To calculate the value of **B** *or of* **B**$_{\prime\prime}$ *from the traction.*

If P denote the pull, and A the area over which it is exerted, the following formulæ (derived from Maxwell's law) may be used:

$$\mathbf{B} = 4{,}965 \sqrt{\frac{P \text{ kilos.}}{A \text{ sq. cm.}}};$$

$$\mathbf{B} = 1{,}316.6 \sqrt{\frac{P \text{ lbs.}}{A \text{ sq. in.}}}; \text{ or}$$

$$\mathbf{B}_{\prime\prime} = 8{,}494 \sqrt{\frac{P \text{ lbs.}}{A \text{ sq. in.}}}.$$

To calculate the requisite cross-section of iron for a given traction.

Reference to p. 89 will show that it is not expedient to attempt to employ tractive forces exceeding 150 pounds per square inch in magnets whose cores are of soft wrought iron, or exceeding 28 pounds per square inch in cast iron. Dividing the given load that is to be sustained by the electromagnet by one or other of these numbers gives the corresponding requisite sectional area of wrought or cast iron respectively.

To calculate the permeability from **B** *or from* **B**$_i$.

This can only be satisfactorily done by referring to a numerical Table (such as Table II. or IV.), or to graphic curves, such as Fig. 18, in which are set down the result of measurements made on actual samples of iron of the quality that is to be used. The values of μ for the two specimens of iron to which Table II. refers may be *approximately* calculated as follows:

For annealed wrought iron, $\mu = \dfrac{17{,}000 - \mathbf{B}}{3.5}$;

For gray cast iron, $\mu = \dfrac{7{,}000 - \mathbf{B}}{3.2}$.

These formulæ must not be used for the wrought iron for tractions that are less than 28 pounds per square inch, nor for cast iron for tractions less than 2¼ pounds per square inch.

To calculate the total magnetic flux which a core of given sectional area can conveniently carry.

It has been shown that it is not expedient to push the magnetization of wrought iron beyond 100,000 lines to the square inch, nor that of cast iron beyond 42,000. These are the highest values that ought to be assumed in designing electromagnets. The total magnetic flux is calculated by multiplying the figure thus assumed by the number of square inches of sectional area.

To calculate the magnetizing power requisite to force a given number of magnetic lines through a definite magnetic reluctance.

Multiply the number which represents the magnetic reluctance by the total number of magnetic lines that are to be forced through it. The product will be the amount of magneto-motive force. If the magnetic reluctance has been expressed on the basis of centimetre measurements, the magneto-motive force, calculated as above, will need to be divided by 1.2566 $\left(i.e., \text{by } \dfrac{4\pi}{10}\right)$ to give the number of ampère turns of requisite magnetizing power. If, however, the magnetic reluctance has been expressed in the units explained below, based upon inch measures, the magnetizing power, calculated by the rule given above, will already be expressed directly in ampère turns.

To calculate the magnetic reluctance of an iron core.

(*a.*) *If dimensions are given in centimetres.*—Magnetic reluctance being directly proportional to length, and inversely proportional to sectional area and to permeability, the following is the formula:

$$\text{Magnetic reluctance} = \frac{l}{A\mu};$$

but the value of μ cannot be inserted until one knows how great **B** is going to be; when reference to Table II. gives μ.

(*b.*) *If dimensions are given in inches.*—In this case we can apply a numerical coefficient, which takes into account the change of units (2.54), and also, at the same time, includes the operation of dividing the magneto-motive force by $\frac{4}{10}$ of π (= 1.2566) to reduce it to ampère turns. So the rule becomes

$$\text{Magnetic reluctance} = \frac{l''}{A''\mu} \times 0.3132.$$

Example.—Find the magnetic reluctance from end to end of a bar of wrought iron 10 inches long, with a cross-section of 4 square inches, on the supposition that the magnetic flux through it will amount to 440,000.

To calculate the total magnetic reluctance of a magnetic circuit.

This is done by calculating the magnetic reluctances of the separate parts, and adding them together. Account must, however, be taken of leakage; for when the flux divides, part going through an armature, part through a leakage path, the law of shunts comes in, and the net reluctance of the joint paths is the reciprocal of the sum of their reciprocals. In the simplest case the magnetic circuit consists of three parts, (1) armature, (2) air in the two gaps, (3) core of the magnet. These three reluctances may be separately written thus:

	For Centimetre Measure.	For Inch Measure.
1. Armature.....	$\dfrac{l_1}{A_1 \mu_1}$	$\dfrac{l''_1}{A''_1 \mu_1} \times 0.3132$
2. The gaps......	$2\dfrac{l_2}{A_2}$	$2\dfrac{l''_2}{A''_2} \times 0.3132$
3. Magnet core...	$\dfrac{l_3}{A_3 \mu_3}$	$\dfrac{l''_3}{A''_3 \mu_3} \times 0.3132$

If the iron used in armature and core is of the same quality, and magnetized up to the same degree of saturation, μ_1 and μ_3 will be alike. For the air-gaps $\mu = 1$, and therefore is not written in.

If there were no leakage, the total reluctance would simply be the sum of these three terms. But when there is leakage, the total reluctance is reduced.

To calculate the ampère turns of magnetizing power requisite to force the desired magnetic flux through the reluctances of the magnetic circuit.

(*a.*) *If dimensions are given in centimetres* the rule is:
Ampère turns = the magnetic flux, multiplied by the magnetic reluctance of the circuit, divided by $\frac{4}{10}$ of π (= 1.2566).

Or, in detail, the three separate amounts of ampère turns required for three principal magnetic reluctances are explained as follows:

Ampère turns required to drive **N** lines through iron of armature $= \mathbf{N} \times \dfrac{l_1}{A_1 \mu_1} \div \dfrac{4\pi}{10}$,

Ampère turns required to drive **N** lines through the two gaps $= \mathbf{N} \times \dfrac{2l_2}{A_2} \div \dfrac{4\pi}{10}$,

Ampère turns required to drive $v\mathbf{N}$ lines through the iron of magnet core $= v\mathbf{N} \times \dfrac{l_3}{A_3 \mu_3} \div \dfrac{4\pi}{10}$,

And, adding up:

Total ampere turns required $= \dfrac{10}{4\pi} \mathbf{N}$ $\left\{ \dfrac{l_1}{A_1 \mu_1} + \dfrac{2l_2}{A_2} + \dfrac{vl_3}{A_3 \mu_3} \right\}$.

(*b.*) *If dimensions are given in inches*, the rule is:

Ampère turns = magnetic flux multiplied by the magnetic reluctance of the circuit.

Or, in detail:

Ampère turns required to drive **N** lines through iron of armature $= \mathbf{N} \times \dfrac{l''_1}{A''_1 \mu_1} \times 0.3132$,

Ampère turns required to drive **N** lines through two gaps $= \mathbf{N} \times \dfrac{2l''_2}{A''_2} \times 0.3132$.

Ampère turns required to drive $v\mathbf{N}$ lines through iron core of magnet $= v\mathbf{N} \times \dfrac{l''_3}{A''_3 \mu_2} \times 0.3132$;

And, adding up:

Total ampere turns required $= 0.3132 \mathbf{N}$ $\left\{ \dfrac{l''_1}{A''_1 \mu_1} + \dfrac{2l''_2}{A''_2} + \dfrac{vl''_3}{A''_3 \mu_2} \right\}$.

It will be noted that here v, the coefficient of allowance for leakage, has been introduced. This has to be calculated as shown later. In the mean time it may be pointed out that, in designing electromagnets for any case where v is approximately known beforehand, the calculation may be simplified by taking the sectional area of the magnet core greater than that of the armature in the same proportion. For example, if it were known that the waste lines that leak were going to be equal in number to those that are usefully employed in the armature (here $v = 2$), the iron of the cores might be made of double the section of that of the armature. In this case μ_3 will approximately equal μ_1.

To calculate the coefficient of allowance for leakage, v.

v = total magnetic flux generated in magnet core ÷ useful magnetic flux through armature. The respective useful and waste magnetic fluxes are proportional to the permeances along their respective paths. *Permeance,* or magnetic conductance, is the reciprocal of the *reluctance,* or magnetic resistance. Call useful permeance through armature and gaps u; and the waste permeance in the stray field w; then

$$v = \frac{u + w}{u}$$

w may be estimated by the Table VIII. or other leakage rules, but should be divided by 2 as the average difference of magnetic potential over the leakage surface is only about half that at the ends of the poles.

Rules for Estimating Magnetic Leakage.

(I. to III. adapted from Prof. Forbes' rules.)

Prop. I. Permeance between two parallel areas facing one another.—Let areas be A_1'' and A_2'' square inches, and distance apart d'' inches, then:

Permeance = $3.193 \times \frac{1}{2}(A''_1 + A''_2) \div d''$.

Prop. II. Permeance between two equal adjacent rectangular areas lying in one plane.—Assuming lines of flow to be semicircles, and that distances d''_1 and d''_2 between their nearest and furthest edges respectively are given, also a'' their width along the parallel edge:

Permeance = $2.274 \times a'' \times \log_{10} \frac{d_2''}{d_1''}$.

Prop. III. Permeance between two equal parallel rectangular areas lying in one plane at some distance apart.—Assume lines of leakage to be quadrants joined by straight lines.

Permeance = $2.274 \times a'' \times \log_{10} \left\{ 1 + \frac{\pi(d''_2 - d''_1)}{d''_1} \right\}$

Prop. IV. Permeance between two equal areas at right angles to one another.

Permeance (if air angle is 90°) = double the respective value calculated by II. or III.

Permeance (if air angle is 270°) = two-thirds times the respective value calculated by II.

If measures are given in centimetres these rules become the following:

I. $\frac{1}{2}(A_1 + A_2) \div d$;

II. $\frac{a}{\pi} \cdot \log_e \frac{d_2}{d_1}$;

III. $\frac{a}{\pi} \log_e \left(1 \times \frac{\pi(d_2 - d_1)}{d_1}\right)$.

Prop. V. Permeance between two parallel cylinders of indefinite length.

The formula for the reluctance is given above: the permeance is the reciprocal of it. Calculations are simplified by reference to Table VIII.

LECTURE III.

SPECIAL DESIGNS.

In continuation of my lecture of last week I have to make a few remarks before entering upon the consideration of special forms of magnets which was to form the entire topic of to-night's lecture. I had not quite finished the experimental results which related to the performance of magnets under various conditions. I had already pointed out that where you require a magnet simply for holding on to its armature common sense (in the form of our simplest formula) dictated that the circuit of iron should be as short as was compatible with getting the required amount of winding upon it. That at once brings us to the question of the difference in performance of long magnets and short ones. Last week we treated that topic so far as this, that if you require your magnet to attract over any range across an air space you require a sufficient amount of exciting power in the circulation of electric current to force the magnetic lines across that resistance, and therefore you require length of core in order to get the required coil wound upon the magnetic circuit. But there is one other way in which the difference of behavior between long and short magnets—I am speaking of horseshoe shapes—comes into play. So far back as 1840, Ritchie found it was more difficult to magnetize steel magnets (using for that purpose electromagnets to stroke them with) if those electromagnets were short than if they were long. He was of course comparing magnets which had the same tractive power, that is to say, presumably had the same section of iron magnetized up to the same degree of magnetization. This difference between long and short cores is obviously to be explained on the same principle as the greater projecting power of the long-legged magnets. In order to force magnetism not only through an iron arch, but through whatever is beyond, which has a lesser permeability for magnetism, whether it be an air-gap or an arch of hard steel destined to retain some of its magnetism, you require magneto-motive force enough to drive the magnetism through that resisting medium; and, therefore, you must have turns of wire. That implies that you must have length of leg on which to wind those turns. Ritchie also found that the amount of magnetism remaining behind in the soft iron arch, after turning off the current at the first removal of the armature, was a little greater with long than with short magnets; and, indeed, it is what we should expect now, knowing the properties of iron, that long pieces, however soft, retain a little more—have a little more memory, as it were, of having been magnetized—than short pieces. Later on I shall have specially to draw your attention to the behavior of short pieces of iron which have no magnetic memory.

WINDING OF THE COPPER.

I now take up the question of winding the copper wire upon the electromagnet. How are we to determine beforehand the amount of wire required and the proper gauge of wire to employ?

The first stage of such a determination is already accomplished; we are already in possession of the formulæ for reckoning out the number of ampère turns of excitation required in any given case. It remains to show how from this to calculate the amount of bobbin space, and the quantity of wire to fill it. Bear in mind that a current of 10 ampères (*i. e.*, as strong as that used for a big arc light) flowing once around the iron produces exactly the same effect magnetically as a current of one ampère flowing around ten times, or as a current of only one-hundredth part of an ampère flowing around a thousand times. In telegraphic work the currents ordinarily used in the lines are quite small, usually from five to twenty thousandths of an ampère; hence in such cases the wire that is wound on need only be a thin one, but it must have a great many turns. Because it is thin and has a great many turns, and is consequently a long wire, it will offer a considerable resistance. That is no advantage, but does not necessarily imply any greater waste of energy than if a thicker coil of fewer turns were used with a correspondingly larger current. Consider a very simple case. Suppose a bobbin is already filled with a certain number of turns of wire, say 100, of a size large enough to carry one ampère, without overheating. It will offer a certain resistance, it will waste a certain amount of the energy of the current, and it will have a certain magnetizing power. Now suppose this bobbin to be rewound with a wire of half the diameter; what will the result be? If the wire is half the diameter it will have one-quarter the sectional area, and the bobbin will hold four times as many turns (assuming insulating materials to occupy the same percentage of the available volume). The current which such a wire will carry will be one-fourth as great. The coil will offer sixteen times as much resistance, being four times as long and of one-fourth the cross-section of the other wire. But the waste of energy will be the same, being proportional to the resistance and to the square of the current; for $16 \times \frac{1}{16} = 1$. Consequently the heating effect will be the same. Also the magnetizing power will be the same, for though the current is only one-quarter of an ampère, it flows around 400 turns; the ampère turns are 100, the same as before. The same argument would hold good with any other numerical instance that might be given. It therefore does not matter in the least to the magnetic behavior of the electromagnet whether it is wound with thick wire or thin wire, provided the thickness of the wire corresponds to the current it has to carry, so that the same number of watts of power are spent in heating it. For a coil wound on a bobbin of given volume the magnetizing power is the same for the same heat waste. But the heat waste increases in a greater ratio than the magnetizing power, if the current in a given coil is increased; for the heat is proportional to the square of the current, and the magnetizing power is simply proportional to the current. Hence it is the heating effect which in reality determines the winding of the wire. We must—assuming that the current will have a certain strength—allow enough volume to admit of our getting the requisite number of ampère turns without overheating. A good way is to assume a current of one ampère while one calculates out the coil. Having done this, the same volume holds good for any other gauge of wire appropriate to any other current. The terms "long coil" magnet and "short coil" magnet are appropriate for those electromagnets which have, respectively, many turns of thin wire and few turns of thick wire. These terms are preferable to "high resistance" and "low resistance," sometimes used to designate the two classes of windings; because, as I have just shown, the resistance of a coil has in itself nothing to do with its magnetizing power. Given the volume occupied by the copper, then for any current density (say, for example, a current density of 2,000 ampères per square inch of cross-section of the copper), the magnetizing power of the coil will be the same for all different gauges of wire. The specific conductivity of the copper itself is of importance; for the better the conductivity the less the heat waste per cubic inch of winding. High conductivity copper is therefore to be preferred in every case.

Now the heat which is thus generated by the current of electricity raises the temperature of the coil (and of the core), and it begins to emit heat from its surface. It may be taken as a sufficient approximation that a single square inch of surface, warmed one degree Fahr.

above the surrounding air, will steadily emit heat at the rate of $\frac{1}{225}$ of a watt. Or, if there is provided only enough surface to allow of a steady emission of heat at the rate of one watt[1] per square inch of surface, the temperature of that surface will rise to about 225 degrees Fahr. above the temperature of the surrounding air. This number is determined by the average emissivity of such substances as cotton, silk, varnish, and other materials of which the surfaces of coils are usually composed.

In the specifications for dynamo machines it is usual to lay down a condition that the coils shall not heat more than a certain number of degrees warmer than the air. With electromagnets it is a safe rule to say that no electromagnet ought ever to heat up to a temperature more than 100 degrees Fahr. above the surrounding air. In many cases it is quite safe to exceed this limit.

The resistance of the insulated copper wire on a bobbin may be approximately calculated by the following rule. If d is the diameter of the naked wire, in mils, and D is the diameter, in mils, of the wire when covered, then the resistance per cubic inch of the coil will be:

$$\text{Ohms per cubic inch} = \frac{960,700}{D^2 \times d^2}$$

---

[1] The *watt* is the unit of rate of expenditure of energy, and is equal to ten million ergs per second, or to 1-746th of a horse power. A current of one ampère, flowing through a resistance of one ohm, spends energy in heating at the rate of one watt. One watt is equivalent to 0.24 calories, per second, of heat. That is to say, the heat developed in one second, by expenditure of energy at the rate of one watt, would suffice to warm one gramme of water through 0.24 (Centigrade) degree. As 252 calories are equal to one British (pound Fahr.) unit of heat, it follows that heat emitted at the rate of one watt would suffice to warm 3.4 pounds of water one degree Fahr. in one hour; or one British unit of heat equals 1,058 watt seconds.

We are therefore able to construct a wire gauge and ampèreage table which will enable us to calculate readily the degree to which a given coil will warm when traversed by a given current, or conversely what volume of coil will be needed to provide the requisite circulation of current without warming beyond any prescribed excess.

Accordingly, I here give a wire-gauge and ampèreage table which we have been using for some time at the Finsbury Technical College. It was calculated out under my instructions by one of the demonstrators of the college, Mr. Eustace Thomas, to whom I am indebted for the great care bestowed upon the calculations

For many purposes, such as for use in telegraphs and electric bells, smaller wires than any of those mentioned in the table are required. The table is, in fact, intended for use in calculating magnets in larger engineering work.

A rough-and-ready rule sometimes given for the size of wire is to allow $\frac{1}{1000}$ square inch per ampère. This is an absurd rule, however, as the figures in the table show. Under the heading 1,000 ampères to square inch, it appears that if a No. 18 S. W. G. wire is used it will at that rate carry 1.81 ampères; that if there is only one layer of wire, it will only warm up 4.64 degrees Fahr., consequently one might wind layer after layer to a depth of 3.3 inches, without getting up to the limit of allowing one square inch per watt for the emission of heat. In very few cases does one want to wind a coil so thick as 3.3 inches. For very few electromagnets is it needful that the layer of coil exceed half an inch in

TABLE X.—WIRE GAUGE AND AMPÈREAGE TABLE. (See notes on opposite page.)

Permissible Ampèreage, Probable Heating, and Permissible Depth.

Nearest B. & S.	S.W.G.	Dimensions.			At 1,000 amps. to sq. inch.			At 2,000 amps. to sq. inch.			At 3,000 amps. to sq. inch.			At 4,000 amps. to sq. inch.		
		Diam. (inch).	Section (sq. inch bare).	Turns to one linear inch (covered).	Turns per sq. inch (covered).	A	F	D	A	F	D	A	F	D	A	F	D
21.	22.	.028	.00062	23.81	624.0	.616	2.28	4.5	1.23	9.12	1.13	1.85	20.52	.50	2.46	36.5	.28
19.	20.	.036	.0010	20.00	440.0	1.018	3.18	3.9	2.036	12.72	.97	3.05	28.62	.43	4.07	50.9	.24
18.	19.	.040	.0012	18.52	377.0	1.26	3.56	3.6	2.52	14.24	.92	3.78	32.04	.41	5.04	57.0	.23
16.	18.	.048	.0018	16.13	296.0	1.81	4.64	3.3	3.62	18.56	.83	5.43	41.70	.37	7.24	74.2	.21
15.	17.	.056	.0024	14.28	224.0	2.4	5.47	3.2	4.8	21.9	.79	7.2	49.2	.35	9.6	87.5	.19
14.	16.	.064	.0032	12.83	181.0	3.2	6.57	3.0	6.4	26.3	.74	9.6	59.1	.33	12.8	105.1	.18
13.	15.	.072	.0040	11.63	149.0	4.0	7.40	2.9	8.0	29.6	.72	12.0	66.6	.32	16.0	118.4	.17
12.	14.	.080	.0050	10.64	124.0	5.0	8.46	2.8	10.0	33.8	.70	15.0	76.3	.31	20.0	135.4	.17
11.	13.	.092	.0060	9.44	98.2	6.6	9.97	2.7	13.2	39.9	.67	19.8	89.7	.30	26.4	159.5	.16
10.	12.	.104	.0085	8.48	79.1	8.5	11.53	2.6	17.0	46.1	.65	25.5	103.8	.29	34.0	184.4	.16
9.	11.	.116	.0105	7.69	65.0	10.5	12.8	2.5	21.0	51.2	.63	31.5	115.2	.28	42.0	204.8	.16
8.	10.	.128	.0128	7.04	54.5	12.8	14.3	2.4	25.6	57.2	.61	38.4	128.7	.27	51.2	228.8	.15
7.	9.	.144	.0163	6.33	44.1	16.3	16.4	2.4	32.6	65.6	.60	48.9	147.6	.27	65.2	262.4	.15
6.	8.	.160	.0201	5.74	36.3	20.1	18.4	2.3	40.2	73.6	.58	60.3	165.6	.26	80.4	294.4	.15
5.	7.	.176	.0243	5.26	30.4	24.3	20.4	2.3	48.6	81.6	.58	72.9	183.6	.26	97.2	326.4	.15
Stranded.																	
7-21	7-22.	0.84	.0043	9.62	101.8	4.3	6.73	4.0	8.6	26.9	.99	12.9	24.6	.44	17.2	107.7	.25
7-19	7-20.	.108	.0072	7.81	67.1	7.13	8.94	3.7	14.3	35.7	.92	21.4	80.5	.44	28.5	143.0	.23
7-16	7-18.	.144	.0128	6.09	40.8	12.7	12.4	3.4	25.4	49.6	.83	38.1	111.6	.39	50.8	198.4	.21
7-14	7-16.	.192	.0229	5.10	28.6	22.9	17.2	3.2	45.8	68.7	.79	68.7	154.5	.35	91.6	274.7	.20
7-13	7-15.	.216	.0249	4.27	20.1	24.9	19.5	3.1	57.8	78.0	.77	86.7	175.4	.34	115.6	311.8	.20
7-12	7-14.	.240	.0356	3.87	16.5	35.6	21.8	3.1	71.2	87.1	.76	106.8	195.9	.34	142.4	348.3	.19
7-11	7-13.	.276	.0462	3.38	12.6	46.2	24.7	3.0	92.4	98.8	.74	138.6	222.3	.33	184.8	395.2	.19
7-10	7-12.	.312	.0595	3.01	9.97	59.5	28.3	2.9	179.0	114.0	.72	178.5	256.5	.32	238.0	456.	.18

thickness; and if the layer is going to be only half an inch thick, or about one-seventh of the 3.3, one may use a current density $\sqrt{7}$ times as great as 1,000 ampères per square inch, without exceeding the limit of safe working. Indeed, with coils only half an inch thick, one may safely employ a current density of 3,000 ampères per square inch, owing to the assistance which the core gives for the dissipation and emission of heat.

Suppose, then, we have designed a horseshoe magnet, with a core one inch in diameter, and that, after considering the work it has to do, it is found that a magnetizing power of 2,400 turns is required; suppose, also,

---

Figures in columns marked A signify number of ampères that the wire carries.

Figures in columns marked F signify number of degrees (Fahrenheit) that the coil will warm up if there is only one layer of wire, and on the assumption that the heat is radiated only from the outer surface of the coil; they are calculated by the following modification of Forbes' rule:

Rise in temperature (Fahrenheit deg.) = 225 × No. of watts lost per sq. inch.
= 159 × sectional area × number of turns to one inch (at 1,000 amps. per sq. inch).

Figures in columns marked D are the depth in inches to which wire may be wound if one watt be lost by each square inch of radiating surface, the outside radiating surface of the bobbin being only considered.

Rule for calculating a 7-strand cable: Diam. of cable = 1.134 × diam. of equivalent round wire.

Figures under heading "Turns to one linear inch" are calculated for cotton covered wires of average thicknesses of coverings used for the different gauges, viz., 14 mils additional diameter on round wires (from No. 22) and 20 mils on stranded or square wire.

Figures under heading "Turns per square inch" are calculated from preceding, allowing 10 per cent. for bedding of layers.

Resistance (ohms) of coil of copper wire, occupying v cubic inches of coil space, and of which the gauge is d mils uncovered, and D mils covered, may be approximately calculated by the rule:

$$\text{ohms} = 960,700 \frac{v}{D^2 d^2}$$

The data respecting sizes of wires of various gauges are kindly furnished by the London Electric Wire Company.

that it is laid down as a condition that the coil must not warm up more than 50 degrees Fahr. above the surrounding air—what volume of coil will be required? Assume, first, that the current will be one ampère; then there will have to be 2,400 turns of a wire which will carry one ampère. If we took a No. 20 S. W. G. wire and wound it to the depth of half an inch, that would give 220 turns per inch length of coil; so that a coil 11 inches long and a little over half an inch deep (or ten layers deep) would give 2,400 turns. Now Table X. shows that if this wire were to carry 1.018 ampères it would heat up 225 degrees Fahr. if wound to a depth of 3.9 inches. If wound to half an inch, it would therefore heat up about 30 degrees Fahr.; and with only one ampère would, of course, heat less. This is too good; try the next thinner wire. No. 22 S. W. G. wire at 2,000 ampères to the square inch will carry 1.23 ampères, and heats 225 degrees if wound up 1.13 inches. If it is only to heat 50 degrees, it must not be wound more than one-fourth inch deep; but if it only carries current of one ampère it may be wound a little deeper—say to 14 layers. There will then be wanted a coil about seven inches long to hold the 2,400 turns. The wire will occupy about 3.85 square inches of total cross-section, and the volume of the space occupied by the winding will be 26.95 cubic inches. Two bobbins, each $3\frac{1}{2}$ inches long and .65 deep, to allow for 14 layers, will be suitable to receive the coils.

By the light of the knowledge one possesses as to the relation between emissivity of surface, rate of heating by current, and limiting temperatures, it is seen how little justification there is for such empirical rules as that which is often given, namely, to make the depth of coil equal to the diameter of the iron core. Consider this in relation to the following fact; that in all those cases where leakage is negligible the number of ampere-turns that will magnetize up a thin core to any prescribed degree of magnetization will magnetize up a core of any section whatever, and of the same length, to the same degree of magnetization. A rule that would increase the depth of copper proportionately to the diameter of the iron core is absurd.

Where less accurate approximations are all that is needed, more simple rules can be given. Here are two cases:

Case 1. *Leakage assumed to be negligible.*—Assume **B** = 16,000, then **H** = 50 (see Table III.). Hence the ampère turns per centimetre of iron will have to be 40, or per inch of iron 102; for **H** is equal to 1.2566 times the ampère turns per centimetre. Now, if the winding is not going to exceed one-half inch in depth, we may allow 4,000 ampères per square inch without serious overheating. And the 4,000 ampère turns will require 2-inch length of coil, or each inch of coil carries 2,000 ampère turns without overheating. Hence each inch of coil one-half inch deep will suffice to magnetize up 20 inches length of iron to the prescribed degree.

Case 2. *Leakage assumed to be 50 per cent.*—Assume **B** in air-gap = **H** = 8,000, then to force this across requires ampère turns 6,400 per centimetre of air, or 16,-250 per inch of air. Now, if winding is not going to exceed one-half inch depth, each inch length of coil will carry 2,000 ampère tu. Hence, eight inches length of coil one-quarter inch deep will be required for one inch length of air, magnetized up to the prescribed degree.

WINDINGS FOR CONSTANT PRESSURE AND FOR CONSTANT CURRENT.

In winding coils for magnets that are to be used on any electric light system, it should be carefully borne in mind that there are separate rules to be considered according to the nature of the supply. If the electric supply is at *constant pressure*, as usual for glow lamps, the winding of coils of electromagnets follows the same rule as the coils of voltmeters. If the supply is with *constant current*, as usual for arc lighting in series, then the coils must be wound with due regard to the current which the wire will carry, when lying in layers of suitable thickness, the number of turns being in this case the same whether thin or thick wire is used.

If we assume that a safe limit of temperature is 90 degrees Fahr. higher than the surrounding air, then the largest current which may be used with a given electromagnet is expressed by the formula:

$$\text{Highest permissible ampères} = 0.63 \sqrt{\frac{s}{r}}$$

where s is the number of square inches of surface of the coils and r their resistance in ohms.

Similarly for coils to be used as shunts we have:

$$\text{Highest permissible volts} = 0.63 \sqrt{sr}$$

The magnetizing power of a coil, supplied at a given number of volts of pressure, is independent of its length, and depends only on its gauge, but the longer the wire the *less* will be the heat waste. On the contrary, when the condition of supply is with a constant number of ampères of current, the magnetizing power of a coil is independent of the gauge of the wire, and depends only on its length; but the larger the gauge the less will be the heat waste.

MISCELLANEOUS RULES ABOUT WINDING.

To reach the same limiting temperature with bobbins of equal size, wound with wires of different gauge, the cross-section of the wire must vary with the current it is to carry; or, in other words, the current density (ampères per square inch) must be the same in each. Table X. shows the ampèreages of the various sizes of wires at four different values of current density.

To raise to the same temperature two similarly shaped coils, differing in size only, and having the gauges of the wires in the same ratio (so that there are the same number of turns on the large coil as on the small one), the currents must be proportional to the square roots of the cubes of the linear dimensions.

Sir William Thomson has given a useful rule for calculating windings of electromagnets of the same type but of different sizes. Similar iron cores, similarly wound with lengths of wire proportional to the squares of their linear dimensions, will, when excited with equal currents, produce equal intensities of magnetic field at points similarly situated with respect to them.

Similar electromagnets of different sizes must have ampère turns proportional to their linear dimensions if they are to be magnetized up to an equal degree of saturation.

It is curious what erroneous notions crop up from time to time about winding electromagnets. In 1869 a certain Mr. Lyttle took out a patent for winding the coils in the following way: Wind the first layer as usual, then bring the wire back to the end where the winding began and wind a second layer, and so on. In this way all the windings will be right-handed, or else all left-handed, not alternately right and left as in the ordinary winding. Lyttle declared that this method of winding a coil gave more powerful effects; so did M. Brisson, who reinvented the same mode of winding in 1873, and solemnly described it. Its alleged superiority was at once disproved by Mr. W. H. Preece, who found the only difference to be that there was more difficulty in carrying out this mode of winding.

Another popular error is that electromagnets in which the wires are badly insulated are more powerful than those in which they are well insulated. This arises from the ignorant use of electromagnets having long, thin coils (of high resistance) with batteries consisting of a few cells (of low electromotive force). In such cases, if some of the coils are short circuited, more current flows, and the magnetizing power may be greater. But the scientific cure is either to rewind the magnet with an appropriate coil of thick wire, or else to apply another battery having an electromotive force that is greater.

SPECIFICATIONS OF ELECTROMAGNETS.

One frequently comes across specifications for construction which prescribe that an electromagnet shall be wound so that its coil shall have a certain resistance. This is an absurdity. Resistance does not help to magnetize the core. A better way of prescribing the winding is to name the ampère turns and the temperature limit of heating. Another way is to prescribe the number of watts of energy which the magnet is to take. Indeed, it would be well if electricians could agree upon some sort of figure of merit by which to compare electromagnets, which should take into account the magnetic output—*i. e.*, the product of magnetic flux into magnetomotive force—the consumption of energy in watts, the temperature rise, and the like.

AMATEUR RULE ABOUT RESISTANCE OF ELECTROMAGNET AND BATTERY.

In dealing with this question of winding copper on a magnet core, I cannot desist from referring to that rule which is so often given, which I often wish might disappear from our text-books—the rule which tells you in effect that you are to waste 50 per cent. of the energy you employ. I refer to the rule which states that you will get the maximum effect out of an electromagnet if you so wind it that the resistance is equal to the resistance of the battery you employ; or that if you have a magnet of a given resistance you ought to employ a battery of the same resistance. What is the meaning of this rule? It is a rule which is absolutely meaningless, unless in the first case the volume of the coil is prescribed once for all, and you cannot alter it; or unless once for all the number of battery elements that you can have is prescribed. If you have to deal with a fixed number of battery elements, and you have to get out of them the biggest effect in your external circuit, and cannot beg, buy, or borrow any more cells, it is perfectly true that, for steady currents, you ought to group them so that their internal resistance is equal to the external resistance that they have to work through; and then, as a matter of fact, half the energy of the battery will be wasted, but the output will be a maximum. Now that is a very nice rule indeed for amateurs, because an amateur generally starts with the notion that he does not want to economize in his rate of working; it does not matter whether the battery is working away furiously, heating itself, and wasting a lot of power; all he wants is to have the biggest possible effect for a little time out of the fewest cells. It is purely an amateur's rule, therefore, about equating the resistance inside to the resistance outside. But it is absolutely fallacious to set up any such rule for serious working; and not only fallacious, but absolutely untrue if you are going to deal with currents that are going to be turned off and on quickly. For any apparatus like an electric bell, or rapid telegraph, or induction coil, or any of those things where the current is going to vary up and down rapidly, it is a false rule, as we shall see presently. What is the real point of view from which one ought to start? I am often asked questions by, shall I say, amateurs, as well as by those who are not amateurs, about prescribing the battery for a given electromagnet, or prescribing an electromagnet for a given battery. Again, I am often told of cases of failure, in which a very little common sense rightly directed might have made a success. What one ought to think about in every case is not the battery, not the electromagnet, but *the line*. If you have a line, then you must have a battery and electromagnet to correspond. If the line is short and thick, a few feet of good copper wire, you should have a short, thick battery, a few big cells or one big cell, and a short, thick coil on your electromagnet. If you have a long, thin line, miles of it, say, you want a long, thin battery (small cells, and a long row of them) and a long, thin coil. That is then our rule: for a short, thick line, a short, thick battery and a short, thick coil; for a long, thin line, a long, thin battery and electromagnet coils to match. You smile; but it is a really good rule that I am giving you; vastly better than the worn-out amateur rule.

But, after all, my rule does not settle the whole question, because there is something more than the whole resistance of the circuit to be taken into account. Whenever you come to rapidly acting apparatus, you have to think of the fact that the current, while varying, is governed not so much by the resistance as by the inertia of the circuit—its electromagnetic inertia. As this is a matter which will claim our especial attention hereafter, I will leave battery rules for the present and proceed with the question of design.

FORMS OF ELECTROMAGNETS.

This at once leads us to consider the classification of forms of magnets. I do not pretend to have found a complete classification. There is a very singular book written by Monsieur Nicklès, in which he classifies under 37 different heads all conceivable kinds of magnets, bidromic, tridromic, monocnemic, multidromic, and I do not know how many more; but the classification is both unmeaning and unmanageable. For my present purpose I will simply pick out those which come under three or four heads, and deal separately with others that do not quite fit under any of the four categories.

Bar Electromagnets.—In the first place there are those which have a straight core, of which there are several specimens on the table here.

Horseshoe Electromagnets.—Then there are the horseshoes, of which some are of one piece, bent, and others here of the more frequent shape, made of three pieces.

Iron-clad Electromagnets.—Then from the horseshoes I go to those magnets in which the return circuit of the iron comes back outside the coil from one end or the other, or from both ends, sometimes in the form of an external tube or jacket, sometimes merely with a parallel return yoke, or two parallel return yokes. All such magnets I propose to call— following the fashion that has been adopted for dynamos—iron-clad electromagnets. One of them, the jacketed electromagnet, is shown in Fig. 12, and there are others not so well known. There is one used by Mr. Cromwell Varley, in which a straight magnet is placed between a couple of iron caps, which fit over the ends, and virtually bring the poles down close together, the circular rim of one cap being the north pole and that of the other cap being the south pole, the two rims being close together. That plan, of course, produces a great tendency to leak across from one rim to the other all round. The advantages, as well as the disadvantages, of the jacketed magnet I alluded to in my last lecture, when I pointed out to you that for all action at a distance it is far better not to have an iron-clad return path, whereas for action in contact the iron-clad magnet was distinctly a very good form. In one form of iron-clad magnet the end of the straight central core is fixed to the middle of a bar of iron, the ends of which are bent up and brought flush

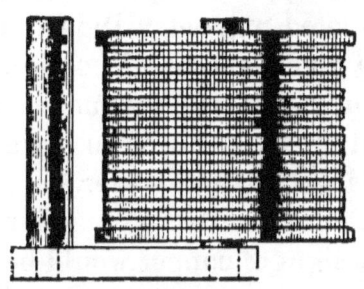

FIG. 50.— CLUB-FOOTED ELECTROMAGNET.

with the top of the bobbin, making thus a tripolar magnet, with one pole between the other two. The armature in this form is a bar which lies right across the three poles. There is an example of this excellent kind of electromagnet applied in one of the forms of electric bell indicator made by Messrs. Gent, of Leicester.

Then besides these three main classes—the straight bar, the horseshoe, and the iron-clad—there is another form which is so useful and so commonly employed in certain work that it deserves to have a name of its own. It is that called by Count Du Moncel the *aimant boiteux*, or club-footed magnet (Fig. 50). It is a horseshoe, in fact, with a coil upon one pole and no coil upon the other. The advantage of that construction is simply, I suppose, that you will save labor—you will only have to wind the wire on one pole instead of two. Whether that is an improvement in any other sense is a question for experiment to determine, but on which theory perhaps might now be able to say something. Count Du Moncel, who made many experiments on this form of magnet, ascertained that there was for an equal weight of copper a slight falling off in power with the club-footed magnet. Indeed, one might almost predict, for a given weight of copper, if you wound all in one coil only, you will not make as many turns as if you wound it in two, the outer turns on the coil being so much larger than the average turn when wound in two coils. Consequently the number of ampère turns with a given weight of copper would be rather smaller, and you would require more current to bring the magnetizing power up to the same value as with the two coils. At the same time the one coil may be produced a little more cheaply than the two; and indeed such electromagnets are really quite common, being largely used for the sake of cheapness and compactness in indicators or electric bells.

Du Moncel tried various experiments about this form to find whether it acted better when the armature was pivoted over one pole or over the other, and found it worked best when the armature was actually hinged on to that pole which comes up through the coil. He made two experiments, trying coils on one or the other limb, the armature being in each case set at an equal distance. In one experiment he found the pull was 35 grammes, with an armature hinged on to the idle pole, and 40 grammes when it was hinged on to the pole which carried the coil.

Another form of electromagnet, having but one coil, is used in the electric bells of church-bell pattern, of which Mr. H. Jensen is the designer. In Jensen's electromagnet a straight cylindrical core receives the bobbin for the coil, and, after this has been pushed into its place, two ovate pole-pieces are screwed upon its ends, serving thus to bring the magnetic circuit across the ends of the bobbin, and forming a magnetic gap along the side of the bobbin. The armature is a rectangular strip of soft iron, about the same length as the core, and is attracted at one end by one pole-piece and at the other end by the other.

EFFECT OF SIZE OF COILS.

Seeing that the magnetizing power which a coil exerts on the magnetic circuit which it surrounds is simply proportional to the ampère turns, it follows that those turns which lie on the outside layers of the coil, though they are further away from the iron core, possess precisely equal magnetizing power. This is strictly true for all closed magnetic circuits; but in those open magnetic circuits where leakage occurs it is only true for those coils which encircle the leakage lines also. For example, in a short bar electromagnet, of the wide turns on the outer layer, those which encircle the middle part of the bar do inclose all the magnetic lines, and are just as operative as the smaller turns that underlie them; while those wide turns which encircle the end portions of the bar are not so efficient, as some of the magnetic lines leak back past these coils.

EFFECT OF POSITION OF COILS.

Among the other researches which Du Moncel made with respect to electromagnets was one on the best position for placing the coil upon the iron core. This is a matter that other experimenters have examined. In Dub's book, "Elektromagnetismus," to which I have several times referred, you will also find many experiments on the best position of a coil; but it is perhaps sufficient to narrate a single example. Du Moncel had four pairs of bobbins made of exactly the same volume, and with 50 metres of wire on each; one pair was 16 centimetres long, another pair eight centimetres, or half the length, with not quite so many turns, because of course the diameter of the outer turn was larger, one four centimetres in length and another two centimetres. These were tried both with bar magnets and horseshoes. It will suffice, perhaps, to give the result of the horseshoe. The horseshoe was made long enough —16 centimetres only, a little over six inches long—to carry the longest coil. Now when the compact coils two centimetres long were used, the pull on the armature at a distance away of two millimetres (it was always the same, of course, in the experiments) was 40 grammes. Using the same weight of wire, but distributed on the coils twice as long, the pull was 55 grammes. Using the coils eight centimetres long it was 75 grammes, and using the coils 16 centimetres long, covering the length of each limb, the pull was 85, clearly showing that, where you have a given length of iron, the best way of winding a magnet to make it pull with its greatest pull is not to heap the coil up against the poles, but to wind it uniformly; for this mode of winding will give you more turns, therefore more ampère turns, therefore more magnetization. An exception might, however, occur in some case where there is a large percentage of leakage. With club-footed magnets results of the same kind are obtained. It was found in every case that it was well to distribute the coil as much as possible along the length of the limb. All these experiments were made with a steady current. It does not follow, however, because winding the wire over the whole length of core is best for steady currents that it is the best winding in the case of a rapidly varying current; indeed, we shall see that it is not.

EFFECT OF SHAPE OF SECTION.

So far as the carrying capacity for magnetic lines is concerned, one shape of section of cores is as good as another; square or rectangular is as good as round if containing equal sectional area. But there are two other reasons, both of which tell in favor of round cores. First, the leakage of magnetic lines from core to core is, for equal mean distances apart, proportional to the surface of the core; and the round core has less surface than square or rectangular of equal section. All edges and corners, moreover, promote leakage. Secondly, the quantity of copper wire that is required for each turn will be less for round cores than for cores any other shape, for of all geometrical figures of equal area the circle is the one of the least periphery.

EFFECT OF DISTANCE BETWEEN POLES.

Another matter that Du Moncel experimented upon, and Dub and Nicklès likewise, was the distance between the poles. Dub considered that it made no difference how far the poles were apart. Nicklès had a special arrangement made which permitted him to move the two upright cores or limbs, nine centimetres high, to and fro on a solid bench or yoke of iron. His armature was 30 centimetres long. Using very weak currents, he found the effect best when the shortest distance between the poles was three centimetres; with a stronger current, 12 centimetres; and with his strongest current, nearly 30 centimetres. I think leakage must have a deal to do with these results. Du Moncel tried various experiments to elucidate this matter, and so did Prof. Hughes in an important but too little known research, which came out in the *Annales Télégraphiques* in the year 1862.

RESEARCHES OF PROFESSOR HUGHES.

His object was to find out the best form of electromagnet, the best distance between the poles, and the best form of armature for the rapid work required in Hughes' printing telegraphs. One word about Hughes' magnet. This diagram (Fig. 51) shows the form of the well known Hughes electromagnet. I feel almost ashamed to say those words "well known," because although on the Continent everybody knows what you mean by a Hughes electromagnet, in England scarcely any one knows what you mean. Englishmen do not even know that Prof. Hughes has invented a special form of electromagnet. Hughes' special form is this: A permanent steel magnet, generally a compound one, having soft iron pole-pieces, and a couple of coils on the pole-pieces only. As I have to speak of Hughes' special contrivance among the mechanisms that will oc-

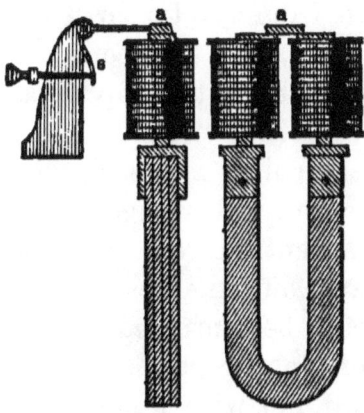

FIG. 51.—HUGHES' ELECTROMAGNET.

cupy our attention next week, I only now refer to this magnet in one particular. If you wish a magnet to work rapidly, you will secure the most rapid action, not when the coils are distributed all along, but when they are heaped up near, not necessarily entirely on, the poles. Hughes made a number of researches to find out what the right length and thickness of these pole-pieces should be. It was found an advantage not to use too thin pole-pieces, otherwise the magnetism from the permanent magnet did not pass through the iron without considerable reluctance, being choked by insufficiency of section; also not to use too thick pieces, otherwise they presented too much surface for leakage across from one to the other. Eventually a particular length was settled upon, in proportion about six times the diameter, or rather longer. In the further researches that Hughes made he used a magnet of shorter form, not shown here, more like those employed in relays, and with an armature from two to three millimetres thick, one centimetre wide, and five centimetres long. The poles were turned over at the top toward one another. Hughes tried whether there was any advantage in making those poles approach one another, and whether there was any advantage in having as long an armature as five centimetres. He tried all different kinds, and plotted out the results of observations in curves, which could be compared and studied. His object was to ascertain the conditions which would give the strongest pull, not with a steady current, but with such currents as were required for operating his printing telegraph instruments; currents which lasted only from one to twenty hundredths of a second. He found it was decidedly an advantage to shorten the length of the armature, so that it did not protrude far over the poles. In fact, he got a sufficient magnetic circuit to secure all the attractive power that he needed, without allowing as much chance of leakage as there would have been had the armature extended a longer distance over the poles. He also tried various forms of armature having very various cross-sections.

POSITION AND FORM OF ARMATURE.

In one of Du Moncel's papers on electromagnets [2] you will also find a discussion on armatures, and the best forms for working in different positions. Among other things in Du Moncel you will find this paradox; that whereas, using a horseshoe magnet with flat poles, and a flat piece of soft iron for armature, it sticks on far tighter when put on edgewise, on the other hand, if you are going to work at a distance, across air, the attraction is far greater when it is set flatwise. I explained the advantage of narrowing the surfaces of contact by the law of traction, B^2 coming in. Why should we have for an action at a distance the greater advantage from placing the armature flatwise to the poles? It is simply that you thereby reduce the reluctance offered by the air-gap to the flow of the magnetic lines. Du Moncel also tried the difference between round armatures and flat ones, and found that a cylindrical armature was only attracted about half as strongly as a prismatic armature having the same surface when at the same distance. Let us examine this fact in the light of the magnetic circuit. The poles are flat. You have at a certain distance away a round armature; there is a certain distance between its nearest side and the polar surfaces. If you have at the same distance away a flat armature having the same surface, and, therefore, about the same tendency to leak, why do you get a greater pull in this case than in that? I think it is clear that, if they are at the same distance away, giving

---

[2] *La Lumière Electrique*, vol. ii.

the same range of motion, there is a greater magnetic reluctance in the case of the round armature, although there is the same periphery, because though the nearest part of the surface is at the prescribed distance, the rest of the under surface is farther away, so that the gain found in substituting an armature with a flat surface is a gain resulting from the diminution in the resistance offered by the air-gap.

POLE-PIECES ON HORSESHOE MAGNETS.

Another of Du Moncel's researches [3] relates to the effect of polar projections or shoes—movable pole-pieces, if you like—upon a horseshoe electromagnet. The core of this magnet was of round iron four centimetres in diameter, and the parallel limbs were ten centimetres long and six centimetres apart. The shoes consisted of two flat pieces of iron slotted out at one end, so that they could be slid along over the poles and brought nearer together. The attraction exerted on a flat armature across air-gaps two millimetres thick was measured by counterpoising. Exciting this electromagnet with a certain battery, it was found that the attraction was greatest when the shoes were pushed to about 15 millimetres, or about one-quarter of the inter-polar distance, apart. The numbers were as follows:

Distance between shoes. Millimetres.	Attraction, in grammes.
2	900
10	1,012
15	1,025
25	965
40	890
60	550

[3] *La Lumière Electrique*, vol. iv., p. 129.

With a stronger battery the magnet without shoes had an attraction of 885 grammes, but with the shoes 15 millimetres apart, 1,195 grammes. When one pole only was employed, the attraction, which was 88 grammes without a shoe, was *diminished* by adding a shoe to 39 grammes!

CONTRAST BETWEEN ELECTROMAGNETS AND PERMANENT MAGNETS.

Now, I want particularly to ask you to guard against the idea that all these results obtained from electromagnets are equally applicable to permanent magnets of steel; they are not, for this simple reason. With an electromagnet, when you put the armature near, and make the magnetic circuit better, you not only get more magnetic lines going through that armature, but you get more magnetic lines going through the whole of the iron. You get more magnetic lines round the bend when you put an armature on to the poles, because you have a magnetic circuit of less reluctance, with the same external magnetizing power in the coils acting around it. Therefore, in that case, you will have a greater magnetic flux all the way round. The data obtained with the electromagnet (Fig. 43), with the exploring coil C on the bend of the core, when the armature was in contact and when it was removed, are most significant. When the armature was present it multiplied the total magnetic flow tenfold for weak currents and nearly threefold for strong currents. But with a steel horseshoe, magnetized once for all, the magnetic lines that flow around the bend of the steel are a fixed quantity, and, however much you diminish the reluctance of the magnetic circuit, you do not create or evoke any more. When the armature is away the magnetic lines arch across, not at the ends of the horseshoe only, but from its flanks, the whole of the magnetic lines leaking somehow across the space. When you have put the armature on, these lines, instead of arching out into space as freely as they did, pass for the most part along the steel limbs and through the iron armature. You may still have a considerable amount of leakage, but you have not made one line more go through the bent part. You have absolutely the same number going through the bend with the armature off as with the armature on. You do not add to the total number by reducing the magnetic reluctance, because you are not working under the influence of a constantly impressed magnetizing force. By putting the armature on to a steel horseshoe magnet you only *collect* the magnetic lines, you do not *multiply* them. This is not a matter of conjecture. A group of my students have been making experiments in the following way: They took this large steel horseshoe magnet (Fig. 52), the length of which from end to end through the steel is 42¼ inches. A light narrow frame was constructed, so that it could be slipped on

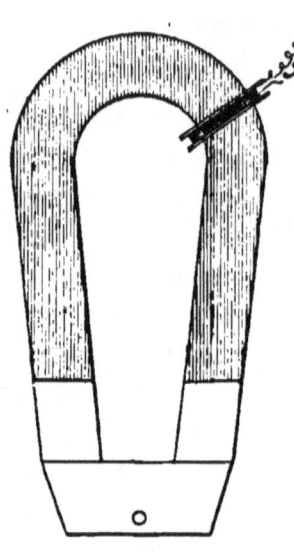

FIG. 52.—EXPERIMENT WITH PERMANENT MAGNET.

over the magnet, and on it were wound 30 turns of fine wire, to serve as an exploring coil. The ends of this coil were carried to a distant part of the laboratory, and connected to a sensitive ballistic galvanometer. The mode of experimenting is as follows: The coil is slipped on over the magnet (or over its armature) to any desired position. The armature of the magnet is placed gently upon the poles, and time enough is allowed to elapse for the galvanometer needle to settle to zero. The armature is then suddenly detached. The first swing measures the change, due to removing the armature, in the number of magnetic lines that pass through the coil in the particular position.

I will roughly repeat the experiment before you; the spot of light on the screen is reflected from my galvanometer at the far end of the table. I place the exploring coil just over the pole, and slide on the armature; then close the galvanometer circuit. Now I detach the armature, and you observe the large swing. I shift the exploring coil, right up to the bend; replace the armature; wait until the spot of light is brought to rest at the zero of the scale. Now, on detaching the armature, the movement of the spot of light is quite imperceptible. In our careful laboratory experiments the effect was noticed inch by inch all along the magnet. The effect when the exploring coil was over the bend was not as great as 1-3000th part of the effect when the coil was hard up to the pole. We are therefore justified in saying that the number of magnetic lines in a permanently magnetized steel horseshoe magnet is not altered by the presence or absence of the armature.

You will have noticed that I always put on the armature gently. It does not do to slam on the armature; every time you do so you knock some of the so-called permanent magnetism out of it. But you may pull off the armature as suddenly as you like. It does the magnet good rather than harm. There is a popular superstition that you ought never to pull off the keeper of a magnet suddenly. On investigation, it is found that the facts are just the other way. You may pull off the keeper as suddenly as you like; but you should never slam it on.

From these experimental results I pass to the special design of electromagnets for special purposes.

ELECTROMAGNETS FOR MAXIMUM TRACTION.

These have already been dealt with in the preceding lecture, the characteristic feature of all the forms suitable for traction being the compact magnetic circuit.

Several times it has been proposed to increase the power of electromagnets by constructing them with intermediate masses of iron between the central core and the outside, between the layers of windings. All these constructions are founded on fallacies. Such iron is far better placed either right inside the coils or right outside them, so that it may properly constitute a part of the magnetic circuit. The constructions known as Camacho's and Cance's, and one patented by Mr. S. A. Varley in 1877, belonging to this delusive order of ideas, are now entirely obsolete.

Another construction which is periodically brought forward as a novelty is the use of iron windings of wire or strip in place of copper winding. The lower electric conductivity of iron, as compared with copper, makes such a construction wasteful of exciting power. To apply equal magnetizing power by means of an iron coil implies the expenditure of about six times as many watts as need be expended if the coil is of copper.

ELECTROMAGNETS FOR MAXIMUM RANGE OF ATTRACTION.

We have already laid down the principle which will enable us to design electromagnets to act at a distance. We want our magnet to project, as it were, its force across the greatest length of air-gap. Clearly, then, such a magnet must have a very large magnetizing power, with many ampère turns upon it, to be able to make the required number of magnetic lines pass across the air resistance. Also it is clear that the poles must not be too close together for its work, otherwise the magnetic lines at one pole will be likely to coil round and take short cuts to the other pole. There must be a wider width between the poles than is desirable in electromagnets for traction.

ELECTROMAGNETS OF MINIMUM WEIGHT.

In designing an apparatus to put on board a boat or a balloon, where weight is a consideration of primary importance, there is again a difference. There are three things that come into play—iron, copper, and electric current. The current weighs nothing; therefore if you are going to sacrifice everything else to weight, you may have comparatively little iron; but you must have enough copper to be able to carry the electric current; and under such circumstances you must not mind heating your wires nearly red hot to pass the biggest possible current. Provide as little copper as you conveniently can, sacrificing economy in that case to the attainment of your object; but, of course, you must use fire-proof material, such as asbestos, for insulating, instead of cotton or silk.

A USEFUL GUIDING PRINCIPLE.

In all cases of design there is one leading principle which will be found of great assistance; namely, that a magnet always tends so to act as though it tried to diminish the length of its magnetic circuit. It tries to grow more compact. This is the reverse of that which holds good with an electric current. The electric circuit always tries to enlarge itself, so as to inclose as much space as possible, but the magnetic circuit always tries to make itself as compact as possible. Armatures are drawn in as near as can be, to close up the magnetic circuit. Many two-pole electromagnets show a tendency to bend together when the current is turned on. One form in particular, which was devised by Ruhmkorff for the purpose of repeating Faraday's celebrated experiment on the magnetic rotation of polarized light, is liable to this defect. Indeed, this form of electromagnet is often designed very badly, the yoke being too thin, both mechanically and magnetically, for the purpose which it has to fulfill.

Here is a small electric bell, constructed by Wagener, of Wiesbaden, the construction of which illustrates this principle. The electromagnet, a horseshoe, lies horizontally; its poles are provided with protruding, curved pins of brass. Through the armature are drilled two holes, so that it can be hung upon the two brass pins, and when so hung up it touches the ends of the iron cores just at one edge, being held from more perfect contact by a spring. There is no complete gap, therefore, in the magnetic circuit. When the current comes and applies a magnetizing power it finds the magnetic

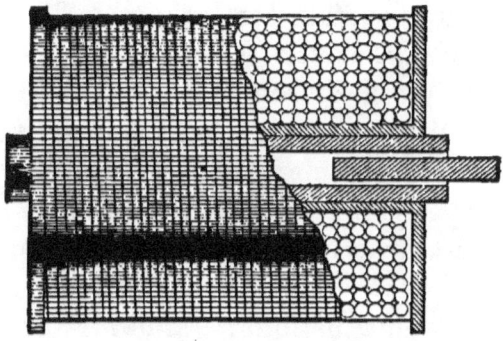

FIG. 53.—ELECTROMAGNETIC POP-GUN.

circuit already complete in the sense that there are no absolute gaps. But the circuit can be bettered by tilting the armature to bring it flat against the polar ends, that being indeed the mode of motion. This is a most reliable and sensitive pattern of bell.

Electromagnetic Pop-Gun.— Here is another curious illustration of the tendency to complete the magnetic circuit. Here is a tubular electromagnet (Fig. 53), consisting of a small bobbin, the core of which is an iron tube about two inches long. There is nothing very unusual about it; it will stick on, as you see, to pieces of iron when the current is turned on. It clearly is an ordinary electromagnet in that respect. Now, suppose I take a little round rod of iron, about an inch long, and put it into the end of the tube, what will happen when I turn on my current? In this apparatus as it stands the magnetic circuit consists of a short length of iron, and then all the rest is air. The magnetic circuit will try to complete itself, not by shortening the iron, but by *lengthening* it; by pushing the piece of iron out so as to afford more surface for leakage. That is exactly what happens; for, as you see, when I turn on the current the little piece of iron shoots out and drops down. You see that little piece of iron shoot out with considerable force. It becomes a sort of magnetic pop-gun. This is an experiment which has been twice discovered. I found it first described by Count Du Moncel, in the pages of *La Lumière Electrique*, under the name of the "pistolet électromagnétique;" and Mr. Shelford Bidwell invented it independently. I am indebted to him for the use of this apparatus. He gave an account of it to the Physical Society in 1885, but the reporter missed it, I suppose, as there is no record in the society's proceedings.

ELECTROMAGNETS FOR USE WITH ALTERNATING CURRENTS.

When you are designing electromagnets for use with alternating currents, it is necessary to make a change in one respect, namely, you must so laminate the iron that internal eddy currents shall not occur; indeed, for all rapid acting electromagnetic apparatus it is a good rule that the iron must not be solid. It is not usual with telegraphic instruments to laminate them by making up the core of bundles of iron plates or wires, but they are often made with tubular cores; that is to say, the cylindrical iron core is drilled with a hole down the middle, and the tube so formed is slit with a saw-cut to prevent the circulation of currents in the substance of the tube. Now, when electromagnets are to be employed with rapidly alternating currents, such as are used for electric lighting, the frequency of the alternations being usually about 100 periods per second, slitting the cores is insufficient to guard against eddy currents; nothing short of completely laminating the cores is a satisfactory remedy. I have here, thanks to the Brush Electric Engineering Company, an electromagnet of the special form that is used in the Brush arc lamp when required for the purpose of working in an alternating current circuit. It has two bobbins that are screwed up against the top of an iron box at the head of the lamp. The iron slab serves as a kind of yoke to carry the magnetism across the top. There are no fixed cores in the bobbins, which are entered by the ends of a pair of yoked plungers. Now in the ordinary Brush lamp for use with a steady current the plungers are simply two round pieces of iron tapped into a common yoke; but for alternate current working this construction must not be used, and instead a U-shaped double plunger is used, made up of laminated iron, riveted together. Of course it is no novelty to use a laminated core; that device, first used by Joule, and then by Cowper, has been repatented rather too often during the past 50 years to be considered as a recent invention.

The alternate rapid reversals of the magnetism in the magnetic field of an electromagnet, when excited by alternating electric currents, sets up eddy currents in every piece of undivided metal within range. All frames, bobbin tubes, bobbin ends and the like must be most carefully slit, otherwise they will overheat. If a domestic flat-iron is placed on the top of the poles of a properly laminated electromagnet, supplied with alternating currents, the flat-iron is speedily heated up by the eddy currents that are generated internally within it. The eddy currents set up by induction in neighboring masses of metal, especially in good conducting metals, such as copper, give rise to many curious phenomena. For example, a copper disc or copper ring placed over the pole of a straight electromagnet so excited is violently repelled. These remarkable phenomena have been recently investigated by Prof. Elihu Thomson, with whose beautiful and elaborate researches we have lately been made conversant in the pages of the technical journals. He rightly attributes many of the repulsion phenomena to the lag in phase of the alternating currents thus induced in the conducting metal. The electromagnetic inertia, or self-inductive property of the electric circuit, causes the currents to rise and fall later in time than the electromotive forces by which they are occasioned. In all such cases the impedance which the circuit offers is made up of two things—resistance and inductance. Both these causes tend to diminish the amount of current that flows, and the inductance also tends to delay the flow.

ELECTROMAGNETS FOR QUICKEST ACTION.

I have already mentioned Hughes' researches on the form of electromagnet best adapted for rapid signaling. I have also incidentally mentioned the fact that where rapidly varying currents are employed, the strength of the electric current that a given battery can yield is determined not so much by the resistance of the electric circuit, but by its electric inertia. It is not a very easy task to explain precisely what happens to an electric circuit when the current is turned on suddenly. The current does not suddenly rise to its full value, being retarded by inertia. The ordinary law of Ohm in its simple form no longer applies; one needs to apply that other law which bears the name of the law of Helmholtz, the use of which is to give us an expression, not for the final value of the current, but for its value at any short time, t, after the current has been turned on. The strength of the current after a lapse of a short time, t, cannot be calculated by the simple process of taking the electromotive force and dividing it by the resistance, as you would calculate steady currents.

In symbols, Helmholtz's law is:

$$i_t = \frac{E}{R}\left(1 - e^{-\frac{R}{L}t}\right)$$

In this formula i_t means the strength of the current after the lapse of a short time t; E is the electromotive force; R the resistance of the whole circuit; L its coefficient of self-induction; and e the number 2.7183, which is the base of the Napierian logarithms. Let us look at this formula; in its general form it resembles Ohm's law, but with a new factor, namely, the expression contained within the brackets. This factor is necessarily a fractional quantity, for it consists of unity less a certain negative exponential, which we will presently further consider. If the factor within brackets is a quantity less than unity, that signifies that i_t will be less than $E \div R$. Now the exponential of negative sign, and with negative fractional index, is rather a troublesome thing to deal with in a popular lecture. Our best way is to calculate some values, and then plot it out as a curve. When once you have got it into the form of a curve, you can begin to think about it, for the curve gives you a mental picture of the facts that the long formula expresses in the abstract. Accordingly we will take the following case: Let $E = 10$ volts; $R = 1$ ohm; and let us take a relatively large self-induction, so as to exaggerate the effect; say let $L = 10$ quads. This gives us the following:

$t_{(sec)}$	$e^{+\frac{R}{L}t}$	i_t
0	1	0
1	1.105	0.950
2	1.221	1.810
5	1.649	3.936
10	2.718	6.343
20	7.389	8.646
30	20.08	9.501
60	403.4	9.975
120	162800.0	9.999

In this case the value of the steady current as calculated by Ohm's law is 10 ampères; but Helmholtz's law shows us that with the great self-induction, which we have assumed to be present, the current, even at the end of 30 seconds, has only risen up to within 95 per cent. of its final value; and only at the end of two minutes has practically attained full strength. These values are set out in the highest curve in Fig. 54, in which, however, the further supposition is made that the number of spirals S in the coils of the electromagnet is 100, so that when the current attains its full value of 10

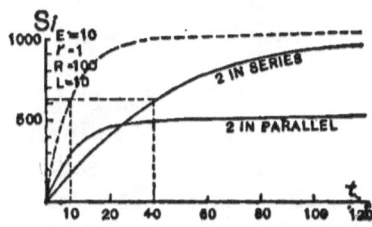

Fig. 54.—Curves of Rise of Currents.

ampères the full magnetizing power will be $Si = 1,000$. It will be noticed that the curve rises from zero at first steeply and nearly in a straight line, then bends over, and then becomes nearly straight again as it gradually rises to its limiting value. The first part of the curve—that relating to the strength of the current after a *very small* interval of time—is the period within which the strength of the current is governed by inertia (*i. e.*, the self-induction) rather than by resistance. In reality the current is not governed either by the self-induction or by the resistance alone, but by the ratio of the two. This ratio is sometimes called the "time-constant" of the circuit, for it represents *the time* which the current takes in that circuit to rise to a definite fraction of its final value. This definite fraction is the fraction $\dfrac{e-1}{e}$; or in decimals, 0.634. All curves of rise of current are alike in general shape—they differ only in scale; that is to say, they differ only in the height to which they will ultimately rise, and in the time they will take to attain this fraction of their final value.

Example (1).—Suppose $E = 10$; $R = 400$ ohms; $L = 8$. The final value of the current will be 0.025 ampère or 25 milliampères. Then the time-constant will be $8 \div 400 =$ 1-50th second.

Example (2).—The P. O. Standard "A" relay has $R = 400$ ohms; $L = 3.25$. It works with 0.5 milliampère current, and therefore will work with 5 Daniell cells through a line of 9,600 ohms. Under these circumstances the time-constant of the instrument on short circuit is 0.0081 second.

It will be noted that the time-constant of a circuit can be reduced either by diminishing the self-induction, or by increasing the resistance. In Fig. 54 the position of the time-constant for the top curve is shown by the vertical dotted line at 10 seconds. The current will take 10 seconds to rise to 0.634 of its final value. This retardation of the rise of current is simply due to the presence of coils and electromagnets in the circuit; the current as it grows being retarded because it has to create magnetic fields in these coils, and so sets up opposing electromotive forces that prevent it from growing all at once to its full strength. Many electricians unacquainted with Helmholtz's law have been in the habit of accounting for this by saying that there is a lag in the iron of the electromagnet cores. They tell you that an iron core cannot be magnetized suddenly; that it takes time to acquire its magnetism. They think it is one of the properties of iron. But we know that the only true time-lag in the magnetization of iron— that which is properly termed " viscous hysteresis "— does not amount to three per cent. of the whole amount of magnetization, takes comparatively a long time to show itself, and cannot therefore be the cause of the retardation which we are considering. There are also electricians who will tell you that when magnetization is suddenly evoked in an iron bar there are induction currents set up in the iron which oppose and delay its magnetization. That they oppose the magnetization is perfectly true; but if you carefully laminate the iron so as to eliminate eddy currents, you will find, strangely enough, that the magnetism rises still more slowly to its final value. For by laminating the iron you have virtually increased the self-inductive action, and increased the time-constant of the circuit, so that the currents rise more slowly than before. The lag is not in the iron, but in the magnetizing current, and the current being retarded, the magnetization is, of course, retarded also.

CONNECTING COILS FOR QUICKEST ACTION.

Now let us apply these most important though rather intricate considerations to the practical problems of the quick working of the electromagnet. Take the case of an electromagnet forming some part of the receiving apparatus of a telegraph system, in which it is desired to secure very rapid working. Suppose the two coils that are wound upon the horseshoe core are connected together in series. The coefficient of self-induction for these two is four times as great as that of either separately; coefficients of self-induction being proportional to the square of the number of turns of wire that surround a given core. Now if the two coils, instead of being put in series, are put in parallel, the coefficient of self-induction will be reduced to the same value as if there were only one coil, because half the line current (which is practically unaltered) will go through each coil. Hence the time-constant of the circuit when the coils are in parallel will be a quarter of that which it is when the coils are in series; on the other hand, for a given line current, the final magnetizing power of the two coils in parallel is only half what it would be with the coils in series. The two lower curves in Fig. 54 illustrate this, from which it is at once plain that the magnetizing power for very brief currents is greater when the two coils are put in parallel with one another than when they are joined in series.

Now this circumstance has been known for some time to telegraph engineers. It has been patented several times over. It has formed the theme of scientific papers which have been read both in France and in England. The explanation generally given of the advantage of uniting the coils in parallel is, I think, fallacious; namely, that the "extra currents" (*i. e.*, currents due to self-induction) set up in the two coils are induced in such directions as tend to help one another when the

LECTURES ON THE ELECTROMAGNET. 215

coils are in series, and to neutralize one another when they are in parallel. It is a fallacy, because in neither case do they neutralize one another. Whichever way the current flows to make the magnetism, it is opposed in the coils while the current is falling by the so-called extra currents. If the current is rising in both coils at the same moment, then, whether the coils are in series or in parallel, the effect of self-induction is to retard the rise of the current. The advantage of parallel grouping is simply that it reduces the time-constant.

BATTERY GROUPING FOR QUICKEST ACTION.

One may consider the question of grouping the battery cells from the same point of view. How does the need for rapid working and the question of time-constant affect the best mode of grouping the battery cells? The amateur's rule, which tells you to so arrange your battery that its internal resistance should be equal to the external resistance, gives you a result wholly wrong for rapid working. The supposed best arrangement will not give you (at the expense even of economy) the best result that might be got out of the given number of cells. Let us take an example and calculate it out, and place the results graphically before our eyes in the form of curves. Suppose the line and electromagnet have together a resistance of six ohms, and that we have 24 small Daniell's cells, each of electromotive force, say, one volt, and of internal resistance four ohms. Also let the coefficient of self-induction of the electromagnet and circuit be six quadrants. When all the cells are in series, the resistance of the battery will be 96 ohms, the total resistance of the circuit 102 ohms, and the full value of the current 0.235 ampère. When all the cells are in parallel the resistance of the battery will be 0.133 ohm, the total resistance 6.133 ohms, and the full value of the current 0.162 ampère. According to the amateur rule of grouping cells so that internal resistance equals external, we must arrange the cells in four parallels, each having six cells in series, so that the internal resistance of the battery will be six ohms, total resistance of circuit 12 ohms, full value of current 0.5 ampère.

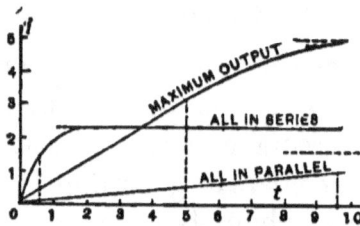

Fig. 55.—Curves of Rise of Current with Different Groupings of Battery.

Now the corresponding time-constants of the circuit in the three cases (calculated by dividing the coefficient of self-induction by the total resistance) will be respectively—in series, 0.06 sec.; in parallel, 0.96 sec.; grouped for maximum steady current, 0.5 sec. From these data we may now draw the three curves, as in Fig. 55, wherein the abscissæ are the values of time in seconds, and the ordinates the current. The faint vertical dotted lines mark the time-constants in the three cases. It will be seen that when rapid working is required the magnetizing current will rise, during short intervals of time, more rapidly if all the cells are put in series than it will do if the cells are grouped according to the amateur rule.

When they are all put in series, so that the battery has a much greater resistance than the rest of the circuit, the current rises much more rapidly, because of the smallness of the time-constant, although it never attains the same ultimate maximum as when grouped in the other way. That is to say, if there is self-induction as well as resistance in the circuit, the amateur rule does not tell you the best way of arranging the battery. There is another mode of regarding the matter which is helpful. Self-induction, while the current is growing, acts as if there were a sort of spurious addition to the resistance of the circuit; and while the current is dying away it acts of course in the other way, as if there were a subtraction from the resistance. Therefore you ought to arrange the batteries so that the internal resistance is equal to the real resistance of the circuit, plus the spurious resistance during that time. But how much is the spurious resistance during that time? It is a resistance proportional to the time that has elapsed since the current was turned on. So then it comes to the question of the length of time for which you want to work it. What fraction of a second do you require your signal to be given in? What is the rate of the vibrator of your electric bell? Suppose you have settled that point, and that the short time during which the current is required to rise is called t; then the apparent resistance at time t after the current is turned on is given by the formula:

$$R_t = R \times e^{\frac{R}{L}t} \div \left(e^{\frac{R}{L}t} - 1\right).$$

TIME-CONSTANTS OF ELECTROMAGNETS.

I may here refer to some determinations made by M. Vaschy,[4] respecting the coefficients of self-induction of the electromagnets of a number of pieces of telegraphic apparatus. Of these I must only quote one result, which is very significant; it relates to the electromagnet of a Morse receiver of the pattern habitually used on the French telegraph lines.

	L, in quadrants.
Bobbins, separately, without iron cores	0.233 and 0.265
Bobbins, separately, with iron cores	1.65 and 1.71
Bobbins, with cores joined by yoke, coils in series	6.37
Bobbins, with armature resting on poles	10.68

It is interesting to note how the perfecting of the magnetic circuit increases the self-induction.

Thanks to the kindness of Mr. Preece, I have been furnished with some most valuable information about the coefficients of self-induction, and the resistance of the standard pattern of relays and other instruments which are used in the British postal telegraph service, from which data one is able to say exactly what the time-constants of those instruments will be on a given circuit, and how long in their case the current will take to rise to any given fraction of its final value. Here let me refer to a very capital paper by Mr. Preece in an old number of the "Journal of the Society of Telegraph Engineers," a paper "On Shunts," in which he treats this question, not as perfectly as it could now be treated

---

[4] "Bulletin de la Société Internationale des Electriciens," 1886.

with the fuller knowledge we have in 1890 about the coefficients of self-induction, but in a very useful and practical way. He showed most completely that the more perfect the magnetic circuit is—though, of course, you are getting more magnetism from your current—the more is that current retarded. Mr. Preece's mode of experiment was extremely simple; he observed the throw of the galvanometer, when the circuit which contained the battery and the electromagnet was opened by a key which at the same moment connected the electro-

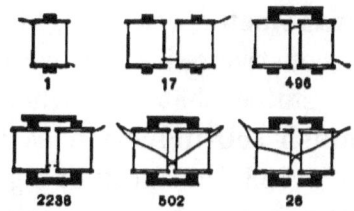

FIG. 56.—ELECTROMAGNETS OF RELAY, AND THEIR EFFECTS.

magnet wires to the galvanometer. The throw of the galvanometer was assumed to represent the extra current which flowed out. Fig. 56 represents a few of the results of Mr. Preece's paper. Take from an ordinary relay a coil, with its iron core, half the electromagnet, so to speak, without any yoke or armature. Connect it up as described, and observe the throw given to the galvanometer. The amount of throw obtained from the single coil was taken as unity, and all others were compared with it. If you join up two such coils as they are usually joined, in series, but without any iron yoke across the cores, the throw was 17. Putting the iron yoke across the cores, to constitute a horseshoe form, 496 was the throw; that is to say, the tendency of this electromagnet to retard the current was 496 times as great as that of the simple coil. But when an armature was put over the top the effect ran up to 2,238. By the mere device of putting the coils in parallel, instead of in series, the 2,238 came down to 502, a little less than the quarter value which would have been expected. Lastly, when the armature and yoke were both of them split in the middle, as is done in fact in all the standard patterns of the British Postal Telegraph relays, the throw of the galvanometer was brought down from 502 to 26. Relays so constructed will work excessively rapidly. Mr. Preece states that with the old pattern of relay having so much self-induction as to give a galvanometer throw of 1,688, the speed of signaling was only from 50 to 60 words per minute; whereas with the standard relays constructed on the new plan, the speed of signaling is from 400 to 450 words per minute. It is a very interesting and beautiful result to arrive at from the experimental study of these magnetic circuits.

SHORT CORES VERSUS LONG CORES.

In considering the forms that are best for rapid action, it ought to be mentioned that the effects of hysteresis in retarding changes in the magnetization of iron cores are much more noticeable in the case of nearly closed magnetic circuits than in short pieces. Electromagnets with iron armatures in contact across their poles will retain, after the current has been cut off, a very large part of their magnetism, even if the cores be of the softest of iron. But so soon as the armature is wrenched off the magnetism disappears. An air-gap in a magnetic circuit always tends to hasten demagnetizing. A magnetic circuit composed of a long air path and a short iron path demagnetizes itself much more rapidly than one composed of a short air path and a long iron path. In long pieces of iron the mutual actions of the various parts tend to keep in them any magnetization that they may possess; hence they are less readily demagnetized. In short pieces where these mutual actions are feeble, or almost absent, the magnetization is less stable and disappears almost instantly on the cessation of the magnetizing force. Short bits and small spheres of iron have no "magnetic memory." Hence the cause of the commonly received opinion among telegraph engineers that for rapid work electromagnets must have short cores. As we have seen, the only reason for employing long cores is to afford the requisite length for winding the wire which is necessary for carrying the needful circulation of current to force the magnetism across the air-gaps. If, for the sake of rapidity of action, length has to be sacrificed, then the coils must be heaped up more thickly on the short cores. The electromagnets in American patterns of telegraphic apparatus usually have shorter cores and a relatively greater thickness of winding upon them than those of European patterns.

LECTURE IV.

ELECTROMAGNETIC MECHANISM.

The task before me to-night comprises the following matters: First, to speak of that particular variety of the electromagnet in which the iron core, instead of being attached to the coils, is movable, and is attracted into them. Secondly, to speak of the modes of equalizing the pull of electromagnets of various sorts over their range of action. Thirdly, to describe sundry mechanisms which depend on electromagnets. Lastly, to discuss the modes of prevention or diminution of the sparking which is so almost invariably found to accompany the break of circuit when one is using an electromagnet.

THE COIL-AND-PLUNGER.

First, then, let me deal with the apparatus wherein an iron core is attracted into a tubular coil or solenoid, an apparatus which, for the sake of brevity, I take the liberty of naming as the *coil-and-plunger*. Now, from quite early times, from 1822 at any rate, it was known that a coil would attract a piece of iron into it, and that this action resembled somewhat the action of a piston going into a cylinder—resembled it, I mean to say, in possessing an extended range of action. The use of such a device as the coil-and-plunger was even patented in this country in 1846 under the name of "a new electromagnet." Electromagnetic engines, or motors, were made on this plan by Page, and afterward by others, and it became generally known as a distinct device. But even now, if you inquire into the literature of the text-books to know what are the peculiar properties of the coil-and-plunger arrangement, you will find that the books give you next to no information. They are content to deal with the thing in very general terms by saying: Here is a sort of sucking magnet; the core is attracted in. Some books go so far as to tell you that the pull is greatest when the core is about half way in; a statement which is true in one particular case, but false in a great many others. Another book tells you that the pull is greatest at a point one centimetre below the centre of the coil, for plungers of all different lengths —which is quite untrue. Another book tells you that a wide coil pulls less powerfully than a narrow one; a statement which is true for some cases and not for others. The books also give you some approximate rules, which, however, are very little to the point. The reason why this ought to receive much more careful consideration is because in this mechanism of coil-and-plunger we have a real means not only of equalizing, but also of vastly extending the range of the pull of the electromagnet. Let us take a very simple example for the sake of contrasting the range of action of the ordinary electromagnet with the range of action of the coil-and-plunger.

Here are some numbers which are given in a paper with which I have long been familiar, a paper read by the late Mr. Robert Hunt in 1856, before the Institution of Civil Engineers, with that eminent engineer, Robert Stephenson, in the chair. Mr. Hunt described the various types of motors, and spoke of this question of the range of action. He recounted some experiments of his own in which the following was the range of action. There was a horseshoe electromagnet which at distance zero—that is, when its armature was in contact—pulled with a pull of 220 pounds; when the distance was made only $\frac{4}{1000}$th of an inch (4 mils), the pull fell to 90 pounds; and when the distance was increased to 20 mils, $\frac{1}{50}$th of an inch), the pull fell to only 36 pounds. The difference from 220 to 36 was within a range of $\frac{1}{50}$th of an inch. He contrasts this with the results given by another mechanism, not quite the simple coil-and-plunger, but a variety of electromagnet brought out about the year 1845 by a Dane, living in Liverpool, named Hjörth, wherein a sort of hollow, truncated cone of iron (Fig. 57), with coils wound upon it—a hollow electromagnet, in fact—was caused to act on another electromagnet, one being caused to plunge into the other. Now we have no information what the pull was at distance zero with this curious arrangement of

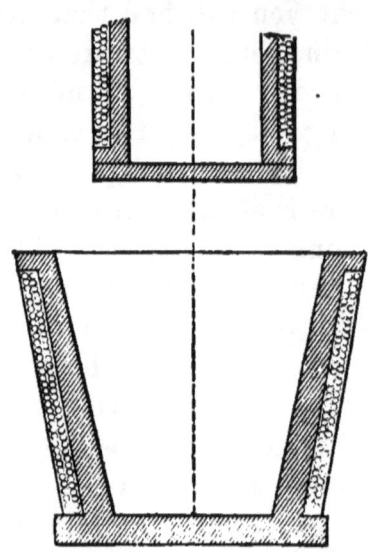

FIG. 57.—HJÖRTH'S ELECTROMAGNETIC MECHANISM.

Hjörth's, but at a distance of one inch the pull (with a very much larger apparatus than Hunt's) was 160 pounds, the pull at three inches was 88 pounds, at five inches 72 pounds. Here, then, we have a range of action going not over $\frac{1}{50}$th of an inch, but over five inches, and falling not from 220 to 36, but from 160 to 72, obviously a much more equable kind of range. At the Institution of Civil Engineers on that occasion a number of the most celebrated men, Joule, Cowper, Sir William Thomson, Mr. Justice Grove, and Prof. Tyndall, discussed these matters—discussed them up and down—from the point of view of range of action, and from the point of view of the fact that there was no means of working them at that time except by the consumption of zinc in a primary battery; and they all came to the conclusion that electric motors would never pay. Robert Stephenson summed up the debate at the end in the following words: "In closing the discussion," he remarked, "there could be no doubt from what had been said that the application of voltaic electricity, in whatever shape it might be developed, was entirely out of the question commercially speaking. Without, however, considering the subject in that point of view, the mechanical applications seemed to involve almost insuperable difficulties. The power exhibited by electromagnetism, though very great, extended through so small a space as to be practically useless. *A powerful magnet might be compared, for the sake of illustration, to a steam engine with an enormous piston but with an exceedingly short stroke; such an arrangement was well known to be very undesirable.*"

Well, from the discussion in 1856—when this question of the length of range was so distinctly set forth—down to the present, there have been a large number of attempts to ascertain exactly how to design a long range electromagnet, and those who have succeeded have, as a general rule, not been the theorists; rather they have been men compelled by force of circumstances to arrive at their result by some kind of—shall we call it—"designing eye," by having a sort of intuitive perception of what was wanted, and going about it in some rough-and-ready way of their own. Indeed, I am afraid had they tried to get much light from calculations based on orthodox notions respecting the surface distribution of magnetism, and all that kind of thing, they would not have been much helped. There is our old friend, the law of inverse squares, which would of course turn up the first thing, and they would be told that it would be impossible to have a magnet that pulled equally through any range, because the pull was certain to vary inversely according to the square of the distance. I noticed that, in a report of my second lecture in one of the London journals, I am announced to have said that the law of inverse squares did not apply to electric forces. I beg to remark I have said no such thing. It is well to be precise as to what one does say. There has been a lively discussion going on quite lately whether sound varies as the square of the distance—or rather, whether the intensity of it does—and the people who dispute on both sides of the case do not seem to know what the law of inverse squares means. I have also seen the statement made last week in the columns of *The*

Times, by one who is supposed to be an eminent authority on eyesight, that the intensity of the color of a scarlet geranium varies inversely with the square of the distance from which you see it. More utter nonsense was never written. The fact is, the law of inverse squares, which is a perfectly true mathematical law, is true not only for electricity, but for light, for sound, and for everything else, provided it is applied to the one case to which a law of inverse squares is applicable. That law is a law expressing the way in which action at a distance falls off when the thing from which the action is proceeding is so small compared with the distance in question that it may be regarded as *a point*. The law of inverse squares is the law universal of action proceeding from a point. The music of an orchestra at 10 feet distance is not four times as loud as at 20 feet distance; for the size of an orchestra cannot be regarded as a mere point in comparison with these distances. If you can conceive of an object giving out a sound, and the object being so small in relation to the distance at which you are away from it that it is a point, the law of inverse squares is all right for that, not for the intensity of your hearing, but for the intensity of that to which your sensation is directed. In no case, however, are sensations absolutely proportional to their causes. When the magnetic action proceeds from something so small that it may be regarded as a point compared with the distance, then the law of inverse squares is necessarily and mathematically true.

You may remember that I produced an apparatus (Fig. 27) which I said was the only apparatus hitherto devised which did directly prove, experimentally, the law of inverse squares for the case of a magnetic pole. There was in it a pole, virtually a point at a considerable distance from a small magnetic needle, which was also virtually a point.

The law of inverse squares is true; but it is not what one works with when one deals with electromagnets having ends of a visible size, acting on armatures themselves of visible sizes, and quite close to them. If you take a case which never occurs in practice, an armature of hard steel, permanently magnetized, so far away from an electromagnet (or rather from one pole only) that the distance between the one pole and the armature on which you are acting is so very great compared with each of them that each of them may be regarded by comparison as a point, then the law of inverse squares may be rightly applied, but not unless.

Now we want to arrive at a true law. We want to know exactly what the law of action of the coil-and-plunger is. It is not a very difficult thing to work out, provided you get hold of the right ideas. We must begin with a simple case, that of a short coil consisting of but one turn, acting on a single point pole. From this we may proceed to consider the effect on a point pole of a long tube of coil. Then we may go on to a more complex case of the tube coil acting on a very long iron core; and last of all from the very long iron core we may pass to the case of a short core.

You all know how a long tube of coil such as this will act on an iron core. Let us make an experiment with it. I turn on the current so that it circulates around the coil along the tube, and when I hold in front of the aperture of the tube this rod of soft iron, it is sucked into the coil. When I pull it out a little way it runs back, as with a spring. The current happens to be a strong one—about 25 ampères; there are about 700 turns of wire on the coil. The rod is about one inch in diameter and 20 inches long. So great is the pull that I cannot pull it entirely out. The pull was very small when the rod was outside, but as soon as it gets in it is pulled actively, runs in and settles down with the ends equally protruding. The tubular coil I have been using is about 14 inches long; but now let us consider a shorter coil. Here is one only half an inch from one end to the other, but I have one somewhere still shorter, so short that the length, parallel to the axis, is very small compared with the diameter of the aperture within. The wire on it consists of but one single turn. Taking such a coil, treating it as only one single ring, with the current going once round, in what way does it act on a magnet that is placed on the axis? First of all, take the case of a very long permanently magnetized steel magnet, so long, indeed, that any action on the more distant pole is so feeble that it may be disregarded altogether and only one pole, say the north pole, is near the coil. In what way will that single turn of coil act on that single pole? This is the rule, that the pull does not vary inversely as the square of the distance, nor as any power at all of the distance measured straight along the axis, but inversely as the cube of the slant distance. Let the point O in Fig. 58 represent the centre of the ring, its radius being y. The line OP

is the axis of the ring, and the distance from O to P we will call x. The slant distance from P to the ring we call a. Then the pull on the axis toward the centre of this coil varies inversely as the cube of a. That law can be plotted out in a curve for the sake of observing the variations of pull at various points along the axis. Allow me to draw your attention to Fig. 59, which represents a section or edge view of the coil. At various distances right and left of the coil are plotted out vertically the corresponding force, the calculations being made for a current of 10 ampères, circulating once around a ring of one centimetre radius. The force with which such a current acts on a magnetic pole of unit strength placed at the central point is 6.28 dynes. If the pole is moved away down the axis, the pull is diminished; at a distance away equal in length to

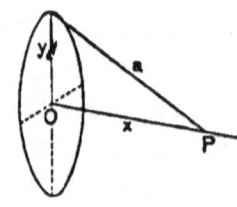

FIG. 58.—ACTION OF SINGLE COIL ON POINT POLE ON AXIS.

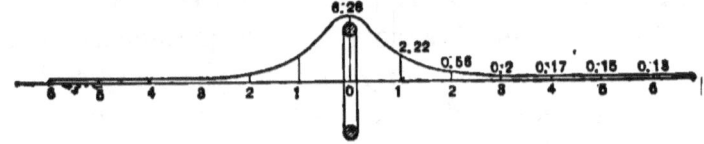

FIG. 59.—ACTION ALONG AXIS OF SINGLE COIL.

the radius it has fallen to 2.22 dynes. At a distance equal to twice the radius, or one diameter, it is only 0.56 dyne, less than one-tenth of what it was at the centre. At two diameters it has fallen to 0.17 dyne, or less than three per cent.; and the force at three diameters is only about two per cent. of that at the centre.

If, then, we could take a *very* long magnet, we may utterly neglect the action on the distant pole. If I had a long steel magnet with the south pole five or six feet away, and the north pole at a point three diameters (*i. e.*, six centimetres in this case) distant from the mouth of the coil, then the pull of the current in one spiral on the north pole three diameters away would be practically negligible; it would be less than two per cent. of what the pull would be of that single coil when the pole was pushed right up into it. But now, in the case of the tubular coil, consisting of at least a whole layer of turns of wire, the action of all of the turns has to be considered. If the nearest of the turns of wire is at a distance equal to three diameters, all the other turns of wire will be at greater distances, and, therefore, if we may neglect such small quantities as two per cent. of the whole amount, we may neglect their action also; for it will be still smaller in amount. Now, for the purpose of arriving at the action of a whole tube of coil, I will adopt a method of plotting devised by Mr. Sayers. Suppose we had a whole tube coiled with copper wire from end to end, its action would be practically the same as though the copper wire were gathered together in small numbers at distant intervals. If, for example, I count the number of turns in a centimetre length of the actual tubular coil, which I used in my first experiment, I find there are four. Now if, instead of having four wires distributed over the centimetre, I had one stout wire in the middle of that space to carry four times the current, the general effect would be the same. This diagram (Fig. 60) is calculated out on the supposition that the effect will be not greatly different if the wires were aggregated in that way, and it is easier to calculate. If, beginning at the end of the tube marked *A*, we take the wires over the first centimetre of length and aggregate them, we can draw a curve, marked 1, for the effect of that lot of wires. For the next lot we could draw a similar curve, but instead of drawing it on the horizontal line we will add the several heights of the second curve on to those of the first, and that gives the curve marked 2; for the third part add the ordinates of another similar curve, and so gradually

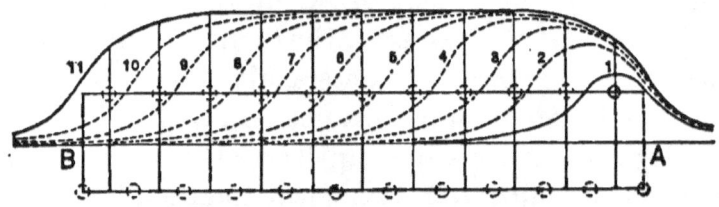

FIG. 60.—ACTION OF TUBULAR COIL.

build up a final curve for the total action of this tubular coil on a unit pole at different points along the axis. This resultant curve begins about 2½ diameters away from the end, rises gently, and then suddenly, and then turns over and becomes nearly flat with a long level back. It does not rise any more after a point about 2½ diameters along from *A*; the curve at that point becomes practically flat, or does not vary more than about one per cent., however long the tube may be. For example, in a tubular coil one inch in diameter and 20 inches long, there will be a uniform magnetic field for about 15 inches along the middle of the coil. In a tubular coil three centimetres in diameter and 40 centimetres long, there will be a uniform magnetic field for about 32 centimetres along the middle of the coil. The meaning of this is that the value of the magnetic forces down the axis of that coil begins outside the mouth of the tube, increases, rises to a certain maximum amount a little within the mouth of the tube, and after that is perfectly constant nearly all the way along the tube, and then falls off symmetrically as you get to the other end. The ordinates drawn to the curve represent the forces at corresponding points along the axis of the tube, and may be taken to represent not simply the magnetizing force, but the pull on a magnetic pole at the end of an indefinitely long, thin steel magnet of fixed strength.

The rule for calculating the intensity of the magnetic force at any point on the axis of the long tubular coil within this region where the force is uniform is: $H = \frac{4}{10}\pi \times$ the ampère turns per centimetre of length. And, as the total magnetizing power of a tubular coil is proportional not only to the intensity of the magnetic force at any point, but also to the length, the integral magnetizing effect on a piece of iron that is inserted into the coil may be taken as practically equal to $\frac{4}{10}\pi \times$ the total number of ampère turns in that portion of the tubular coil which surrounds the iron. If the iron protrudes as much as three diameters at both ends, the total magnetizing force is simply $\frac{4}{10}\pi \times$ the whole number of ampère turns.

Now that case is of course not the one we are usually

dealing with. We cannot procure steel magnets with unalterable poles of fixed strength. Even the hardest steel magnet, magnetized so as to give us a permanent pole near or at the end of it—quite close up to the end of it—when you put it into a magnetizing coil—becomes by that fact further magnetized. Its pole becomes strengthened as it is drawn in, so that the case of an unalterable pole is not one which can actually be realized. One does not usually work with steel; one works with soft iron plungers which are not magnetized at all when at a distance away, but become magnetized in the act of being placed at the mouth of the coil, and which become more highly magnetized the further they go in. They tend, indeed, to settle down, with the ends protruding equally, for that is the position where they most nearly complete the magnetic circuit; where, therefore, they are most completely and highly magnetized. Accordingly we have this fact to deal with, and whatever may be the magnetizing forces all along the tube, the magnetism of the entering core will increase as it goes on. We must therefore have recourse to the following procedure: We will construct a curve in which we will plot not simply the magnetizing forces of the spiral at different points, but the product of the magnetizing forces into the magnetism of the core which itself increases as the core moves in. The curve with a flat top to it corresponds to an ideal case of a single pole of constant strength. We wish to pass from this to a curve which shall represent a real case, with an iron core. Let us still suppose that we are using a very long core, one so long that when the front pole has entered the coil the other end is still a long way off. With an iron core of course it depends on the size and quality of the iron as to how much magnetism you get for a given amount of magnetizing power. When the core has entered up to a certain point you have all the magnetizing forces up to that point acting on it; it acquires a certain amount of magnetism, so that the pull will necessarily go on increasing and increasing, although the intensity of the magnetic force from point to point along

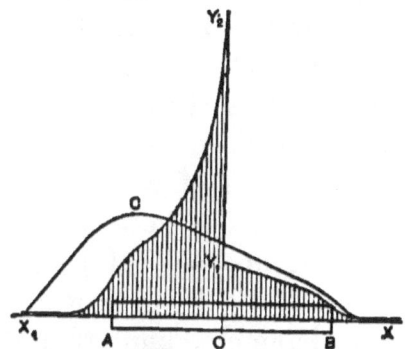

Fig. 61.—Diagram of Force and Work of Coil-and-Plunger.

the axis of the coil remains the same, until within about two diameters from the far end. Although the magnetic force inside the long spiral remains the same, because the magnetism of the core is increasing, the pull goes on increasing and increasing (if the iron does not get saturated) at an almost uniform rate all the way up until the piece of iron has been poked pretty nearly through to the distant end. In Fig. 61 a tubular coil, BA, is represented. Suppose a long iron core is placed on the axis to the right, and that its end is gradually brought up toward B. When it arrives at X the pull becomes sensible, and increases at first rapidly, as the core enters the mouth of the tube, then gently, as the core travels along, attaining a maximum, C, about at the further end, A, of the tube. When it approaches to the other end, A, it comes to the region where the magnetizing force falls off, but the magnetism is still going on increasing, because something is still being added to the total magnetizing power, and these two effects nearly balance one another, so that the pull arrives at the maximum. This is the highest point, C, on the curve; the greatest pull occurring just as the end of the iron core arrives at the bottom or far end of the tubular coil; from which point there is a very rapid falling off. The question of rapidity of descent from that point depends only on how long the core is. If the core is a very long one, so that its other pole is still very far away, you have a long, slow descent going on over some three diameters, and gradually vanishing. If, however, the other pole is coming up within measurable distance of B, then the curve will come down more rapidly to a definite point, X_1. To take a simple case where the iron core is twice as long as the coil, its curve will descend in pretty nearly a straight line down to a point such that the ends of the iron rod stand out equally from the ends of the tube.

Precisely similar effects will occur in all other cases where the plunger is considerably longer than (at least twice as long as) the coil surrounding it. If you take a different case, however, you will get another effect. Take the case of a plunger of the same length as the coil, then this is what necessarily happens. At first the effects are much the same; but as soon as the core has entered about half, or a little more than half, its length you begin to have the action of the other pole that is left protruding outside tending to pull the plunger back; and although the magnetizing force goes on increasing the further the plunger enters, the repulsion exerted by the coil on the other pole of the plunger keeps increasing still faster as this end nears the mouth of the coil. In that case the maximum will occur at a point a little further than half way along the coil, and from that point the curve will descend and go to zero at A; that is to say, there will be no pull when both ends of the plunger coincide with the two ends of the coil. If you take a plunger that is a little shorter than the coil, then you find that the attraction comes down to zero at an earlier period still. The maximum pull occurs earlier, and so does the reduction of the pull to zero; there being no action at all upon the short core when it lies wholly within that region of the tube within which the intensity of the magnetic force is uniform. That is to say, for any portion of this tube corresponding to the flat top of the curve of Fig. 60, if the plunger of iron is so short as to lie wholly within that region, then there is no action upon it; it is not pulled either way. Now these things can be not only predicted by the help of such a law as that, but verified by experiment. Here is a set of tubular coils which we use at the Finsbury Technical College for the purpose of verifying these laws. There is one here about nine inches long, one about half that length, another just a quarter.

They are all made alike in this way, that they have exactly the same weight of copper wire, cut from the same hank, upon them. There are, of course, more turns on the long one than on the shorter, because with the shorter ones each turn requires, on the average, a larger amount of wire, and therefore the same weight of wire will not make the same number of windings. We use that very simple apparatus, a Salter's balance, to measure the pull exerted down to different distances on cores of various lengths. You find in every case the pull increases and becomes a maximum, then diminishes. We will now make the experiment, taking first a long plunger, roughly about twice as long as the coil. The pull increases as the plunger goes down, and the maximum pull occurs just when the lower end gets to the bottom; beyond that the pull is less. Using the same plunger with these shorter coils, one finds the same thing, in fact more marked, for we have now a core which is more than twice the length of the coil. So we find, taking in all these cases, that the maximum pull occurs not when the plunger is half way in, as the books say, but when the bottom end of it is just beginning to come out through the bottom of the coil that we are using. If, however, we take a shorter plunger, the result is different. Here is one just the same length as the coil. With this one the maximum pull does occur when the core is about half way in; the maximum pull is just about at the middle. Again, with a very short core—here is one about one-sixth of the length of the coil—the maximum pull occurs as it is going into the mouth of the coil; and when both ends have gone in so far that it gets into the region of equable magnetic field there is no more pull on one end than on the other; one end is trying to move with a certain force down the tube, and the other end is trying to move with exactly equal force up the tube, and the two balance one another. If we carry that to a still more extreme case, and employ a little round ball of iron to explore down the tube, you will find this curious result, that the only place where any pull occurs on the ball is just as it is going in at the mouth. For about half an inch in the neck of the coil there is a pull; but there is no pull down the interior of the tube at all, and there is no measurable pull outside.

Now these actions of the coil on the core are capable of being viewed from another standpoint. Every engineer knows that the work done by a force has to be measured by multiplying together the force and the distance through which its point of application moves forward. Here we have a varying force acting over a certain range. We ought, therefore, to take the amount of the force at each point, and multiply that by the adjacent little bit of range, averaging the force over that range, and then take the next value of force with the next little bit of range, and so consider in small portions the work done along the whole length of travel. If we call the length of travel x the element of length must be called dx. Multiply that by f, the force. The force multiplied by the element of length gives us the work, dw, done in that short range. Now the whole work over the whole travel is made up of the sum of such elements all added together; that is to say, we have to take all the various values of f, multiply each by its own short range dx, and the sum of all those, writing $\int$ for the sum, would be equal to the sum of all the work; that is to say, the whole work done in putting the thing together will be written:

$$w = \int f\, dx.$$

Now what I want you to think about is this: Here, say, is a coil, and there is a distant core. Though there is a current in the coil, it is so far away from the core that practically there is no action: bring them nearer and nearer together; presently they begin to act on one another; there is a pull, which increases as the core enters, then comes to a maximum, then dies away as the end of the core begins to protrude at the other side. There is no further pull at all when the two ends stand out equally. Now there has been a certain total amount of work done by this apparatus. Every engineer knows that if we can ascertain the force at every point along the line of travel the work done in that travel is readily expressed by the area of the force curve. Think of the curve $X\,C\,X_1$, in Fig. 61, the ordinates of which represent the forces. The whole area underneath this curve represents the work done by the system, and therefore represents equally the work you would have to do upon it in pulling the system apart. The area under the curve represents the total work done in attracting in the iron plunger, with a pull distributed over the range $X\,X_1$.

Now I want you to compare that with the case of an

electromagnet where, instead of having this distributed pull, you have a much stronger pull over a much shorter range. I have endeavored to contrast the two in the other curves drawn in Fig. 61. Suppose we have our coil, and suppose the core, instead of being made of one rod such as this, were made in two parts, so that they could be put together with a screw in the middle, or fastened together in any other mechanical way. Now first treat this rod as a single plunger, screw the two parts together, and begin with the operation of allowing it to enter into the coil; the work done will be the area under the curve which we have already considered. Let us divide the iron core into two. First of all put in one end of it; it will be attracted up in a precisely similar fashion, only, being a shorter bar, the maximum would be a little displaced. Let it be drawn in up to half way only; we have now a tube half filled with iron, and in doing so we shall have had a certain amount of work done by the apparatus. As the piece of iron is shorter, the force curve, which ascends from X to Y_1, will lie a little lower than the curve XCX_1; but the area under that lower curve, which stops half way, will be the work done by the attraction of this half core. Now go to the other end and put in the other half of the iron You now have not only the attraction of the tube, but that of the piece which is already in place, acting like an electromagnet. Beginning with a gentle attraction, it soon runs up, and draws the force curve to a tremendously steep peak, becoming a very great force when the distance asunder is very small. We have therefore in this case a totally different curve made up of two parts, a part for the putting in of the first half of the core, and a steeper part for the second; but the net result is, we have the same quantity of iron magnetized in exactly the same manner by the same quantity of electric current running round the same amount of copper wire—that is to say, the total amount of work done in these two cases is necessarily equal. Whether you allow the entire plunger to come in by a gentle pull over a long range, or whether you put the core in in two pieces—one part with a gentle pull and the other with a sudden spring up at the end—the total work must be the same; that is to say, the total area under our two new curves must be the same as the area under the old curve. The advantage, then, of this coil-and-plunger method of employing iron and copper is, not that it gets any more work out of the same expenditure of energy, but that it distributes the pull over a considerable range. It does not, however, equalize it altogether over the range of travel.

A number of experimental researches have been made from time to time to elucidate the working of the coil-and-plunger. Hankel, in 1850, examined the relation between the pull in a given portion of the plunger and the exciting power. He found that, so long as the iron core was so thick and the exciting power so small that magnetization of the iron never approached saturation, the pull was proportional to the square of the current, and was also proportional to the square of the number of turns of wire. Putting these two facts together, we get the rule—which is true only for an unsaturated core in a given position—that the pull is proportional to the square of the ampère turns. This might have been expected, for the magnetism of the iron core will, under the assumptions made above, be proportional to the ampère turns, and the intensity of the magnetic field in which it is placed being also proportional to the ampère turns, the pull, which is the product of the magnetism and of the intensity of the field, ought to be proportional to the square of the ampère turns.

Dub, who examined cores of different thicknesses, found the attraction to vary as the square root of the diameter of the core. His own experiments show that this is inexact, and that the force is quite as nearly proportional to the diameter as to its square root. There is again reason for this. The magnetic circuit consists largely of air paths by which the magnetic lines flow from one end to the other. As the main part of the magnetic reluctance of the circuit is that of the air, anything which reduces the air reluctance increases the magnetization, and, consequently, the pull. Now, in this case, the reluctance of the air paths is mainly governed by the surface exposed by the end portions of the iron core. Increasing these diminishes the reluctance, and increases the magnetization by a corresponding amount. Von Waltenhofen, in 1870, compared the attraction exerted by two equal (short) tubular coils on two iron cores, one of which was a solid cylindrical rod, and the other a tube of equal length and weight, and found the two to be more powerfully attracted. Doubtless, the effect of the increased service in diminishing the reluctance of the magnetic circuit explains the cause of the observation.

Von Feilitzsch compared the action of a tubular coil upon a plunger of soft iron with that exerted by the same coil upon a core of hard magnetized steel of equal dimensions. The plungers (Fig. 62) were each 10.1 centimetres long, the coil being 29.5 centimetres in length and 4.2 in diameter. The steel magnet showed a maximum attraction when it had plunged to a depth of five centimetres, while the iron core had its maximum at a depth of seven centimetres, doubtless because its own magnetization went on increasing more than did that of the steel core. As the uniform field region began at a depth of about eight centimetres, and the cores were 10.1 centimetres in length, one would expect the attracting force to come to zero when the cores had plunged in to a depth of about 18 centimetres. As a matter of fact, the zero point was reached a little earlier. It will be noticed that the pull at the maximum was a little greater in the case of the iron plunger.

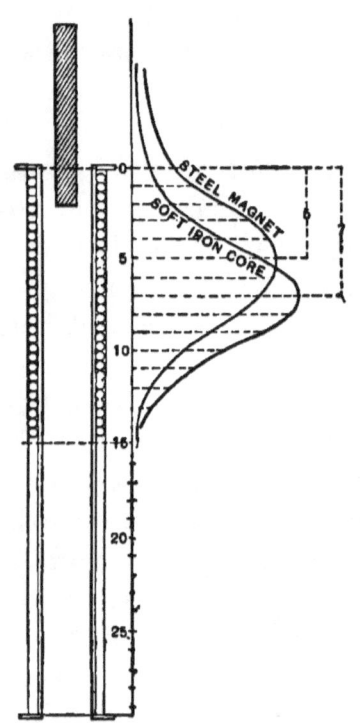

FIG. 62.—VON FEILITZSCH'S EXPERIMENT ON PLUNGERS OF IRON AND STEEL.

The most careful researches of late years are those made by Dr. Theodore Bruger, in 1886. One of his re-

LECTURES ON THE ELECTROMAGNET.   245

searches, in which a cylindrical iron plunger was used, is represented by two of the curves in Fig. 63. He used two coils, one 3½ centimetres long, the other seven centimetres long. These are indicated in the bottom left-hand corner. The exciting current was a little over eight ampères. The cylindrical plunger was 39 centi-

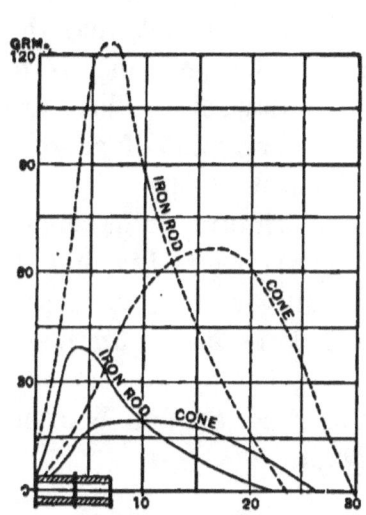

FIG. 63.—BRUGER'S EXPERIMENTS ON COILS AND PLUNGERS.

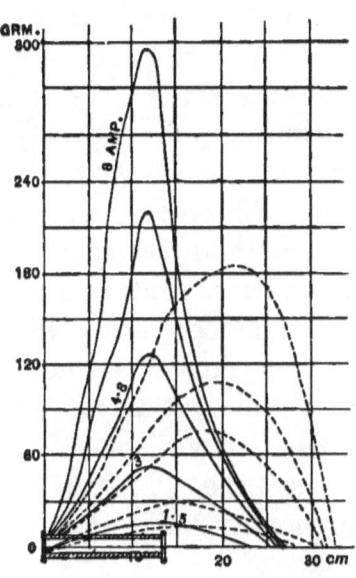

FIG. 64.—BRUGER'S EXPERIMENTS, USING CURRENTS OF VARIOUS STRENGTHS.

metres long. The plunger is supposed, in the diagram, to enter on the left, and the number of grammes of pull is plotted out opposite the position of the entering end of the plunger. As the two curves show by their steep peaks, the maximum pull occurs just when the end of the plunger begins to emerge through the coil, and the pull comes down to zero when the ends of the core protrude equally. In this figure the dotted curves relate to the use of the longer of the two coils. The height of the peak, with the coil of double length, is nearly four times as great, there being double ampère turns of excitation. In some other experiments, which are plotted in Fig. 64, the same core was used with a tubular coil 13 centimetres long. Using currents of various strengths, 1.5 ampère, 3, 4.8, 6, or 8 ampères, the pull is of course different, but broadly, you get the same effect, that the maximum pull occurs just where the pole begins to come out at the far end of the tubular coil. There are slight differences; with the smallest amount of current the maximum is exactly over the end of the tube, but with currents rather larger the maximum point comes a little farther back. When the core gets well saturated, the force curve does not go on rising so far; it begins to turn over at an earlier stage, and the maximum place is necessarily displaced a little way back from the end of the tube. That was also observed by Von Waltenhofen when using the steel magnet.

EFFECT OF USING CONED PLUNGERS.

But now, if, instead of employing a cylindrical core, you employ one that is pointed, you find this completely alters the position of the maximum pull, for now the point is insufficient to carry the whole of the magnetic lines which are formed in the iron rod. They do not come out at the point, but filter through, so to speak, along the sides of the core. The region where the magnetic lines come up through the iron into the air is no longer a definite "pole" at or near the end of the rod, but is distributed over a considerable surface; consequently when the point begins to poke its nose out, you still have a larger portion of iron up the tube, and the pull, instead of coming to a maximum at that position, is distributed over a wider range. I am now making the experiment roughly with my spring balance and a conical plunger, and I think you will be able to notice a marked difference between this case and that of the cylindrical plunger. The pull increases as the plunger enters, but the maximum is not so well defined with a pointed core as it is with one that is flat ended. This essential difference between coned plungers and cylindrical ones was discovered by an engineer of the name of Krizik, who applied his discovery in the mechanism of the Pilsen arc lamps. Coned plungers were also examined by Bruger. In Fig. 63 are given the curves that correspond to the use of a coned iron core, as well as those corresponding to the use of the cylindrical iron rod. You will notice that, as compared with the cylindrical plunger, the coned core never gave so big a pull, and the maximum occurred not as the tip emerged, but when it got a very considerable way out on the other side. So it is with both the shorter and the longer coil. The dotted curves in Fig. 64 represent the behavior of a coned plunger. With the longer coil represented, and with various currents, the maximum pull occurred when the tip had come a considerable way out; and the position of the maximum pull, instead of being brought nearer to the entering end with a high magnetizing current, was actually caused to occur further down. The range of action became extended with large currents as compared with small ones. Bruger also investigated the case of cores of very irregular shapes, resembling, for example, the shank of a screw-driver, and found a very curious and irregular force curve. There is a good deal more yet to be done, I fancy, in examining this question of distributing the pull on an attracted core by altering the shape of it, but Bruger has shown us the way, and we ought not to find very much difficulty in following him.

OTHER MODES OF EXTENDING RANGE OF ACTION.

Another way of altering the distribution of the pull is to alter the distribution of the wire on the coil. Instead of having a coned core use a coned coil, the winding being heaped up thicker at one end than at the other. Such a coil, wound in steps of increasing thickness, has been used for some years by Gaiffe in his arc lamp; it has also been patented in Germany by Leupold. M. Trève has made the suggestion to employ an iron wire coil, so to utilize the magnetism of the iron that is carrying the current. Trève declares that such coils possess for an equal current four times the pulling power. I doubt whether that is so; but even if it were, we must remember that to drive any given current through an iron wire, instead of a copper wire of the same bulk, implies that we must force the current through six times the resistance; and, therefore, we shall have to employ six times the horse power to drive the same current through the iron wire coil, so that there is really no gain. Again, a suggestion has been made to inclose in an iron jacket the coil employed in this way. Iron-clad solenoids have been employed from time to time. But they do not increase the range of action. What they do is to tend to prevent the falling off of the internal pull at the region within the mouth of the coil. It equalizes the internal pull at the expense of all external action. An iron-clad solenoid has practically no attraction at all on anything outside of it, not even on an iron core placed at a distance of half a diameter of the aperture; it is only when the core is inside the tube that the attraction begins, and the magnetizing power is practically uniform from end to end. Last year I wished to make use of this property for some experiments on the action of magnetism on light, and for that purpose I had built, by Messrs. Paterson and Cooper, this powerful coil, which is provided with a tubular iron jacket outside, and a thick iron disc perforated by a central hole covering each end. The magnetic circuit around the exterior of the coil is practically completed with soft iron. With this coil, one may take it, there is an absolutely uniform magnetic field from one end of the tube to the other; not falling off at the ends as it would do if the magnetic circuit had simply an air return. The whole of the ampère turns of exciting power are employed in magnetizing the central space, in which therefore the actions are very powerful and uniform. The coil and its uses were described in my lecture last year at the Royal Institution on "Optical Torque."

MODIFICATIONS OF THE COIL-AND-PLUNGER.

In one variety of the coil-and-plunger mechanism a second coil is wound on the plunger. Hjörth used this modification, and the same thing has been employed in several arc lamps. There is a series of drawings upon this wall depicting the mechanism of about a dozen different forms of arc lamp, all made by Messrs. Paterson and Cooper. In one of these there is a plunger with a coil on it drawn into a tubular coil, the current flowing successively through both coils. In another there are two separate coils in separate circuits, one of thin wire and one of thick, one being connected in series with the arc, and one in shunt.

DIFFERENTIAL COIL-AND-PLUNGER.

There is a third drawing here, showing the arrangement which was originally introduced by Siemens, wherein a plunger is drawn at one end into the coil that is in the main circuit, and at the other end into a coil that is in shunt. That differential arrangement has certain peculiar properties of which I must not now stop to speak in detail. It is obvious that where one core plunges its opposite ends into two coils, the magnetization will depend on both coils, and the resultant pull will not be simply the difference between the pull of the two coils acting each separately. There is, however, another differential arrangement, used in the Brockie-Pell and other arc lamps, in which there are two separate plungers attached to the two ends of a *see-saw* lever. In this case the two magnetizing actions are separate. In a third differential arrangement there is but one plunger and one tubular bobbin, upon which are wound the two coils, differentially, so that the action on the plunger is simply due to the difference between the ampère turns circulating in the two separate wires.

COIL-AND-PLUNGER COIL.

When one abandons iron altogether, and merely uses two tubular coils, one of wide diameter and another of narrower diameter, capable of entering into the former, and passes electric currents through both of them, if the currents are circulating in the same fashion through both of them they will be drawn into one another. This arrangement has also been used in arc lamps. The parallel currents attract one another inversely, not as the square of the distance, but approximately as the distance. This is one of those little details about which it is as well to be clear. About once a year some kind friend from a distance writes to me pointing out a little slip that he says occurs in my book on electricity, in the passage where I am speaking about the attraction of parallel wires. I have made the terrible blunder of leaving out the word square; for I say the attraction varies inversely as the distance, and my readers are kind enough to correct me. Now when I wrote that passage I considered carefully what I had to write, and the attraction does not vary inversely as the square of the distance, because two parallel wires do not act on one another as two points. They act as two straight lines or two parallel lines, and the attraction between two parallel lines of current, or two parallel lines of magnetism, or two parallel lines of anything else that can attract, will not act inversely as the square, but simply inversely as the distance in between.

INTERMEDIATE FORMS.

Now this property of the coil-and-plunger of extending the range of action has been adopted in various ways by inventors whose object was to try and make electromagnets with a sort of intermediate range. For certain purposes it is desirable to construct an electromagnet which, while having the powerful pull of the electromagnet, should have over its limited range of action a more equable pull, resembling in this respect the equalizing of range of the coil-and-plunger. Some of these intermediate forms of apparatus are shown in the following diagrams.

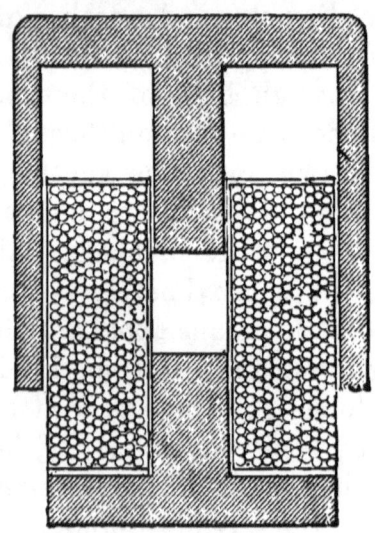

FIG. 65.—PLUNGER ELECTROMAGNET OF STEVENS AND HARDY.

Here (Fig. 65) is a peculiar form of electromagnet; it combines some of the features of the iron-clad electromagnet with those of the movable plunger; it has a limited range of action, but is of great power over that range, owing to its excellent magnetic circuit. It was invented in 1870 by Stevens and Hardy for use in an electric motor for running sewing machines. A very similar form is used in Weston's arc lamp. A form of plunger electromagnet invented by Holroyd Smith in 1877 resembles Fig. 65 inverted, the coil being surrounded by an iron jacket, while a plunger furnished at the top with an iron disc descends down the central tube to meet the iron at the bottom.

Then there is another variety, of which I was able to show an example last week by the kindness of the Brush Company, namely, the plunger electromagnet employed in the Brush arc lamps. A couple of tubular coils receive each an iron plunger, connected together by a yoke; while above, the magnetic circuit is partially completed by the sheet of iron which forms part of the inclosing box. You have here, also, the advantage of

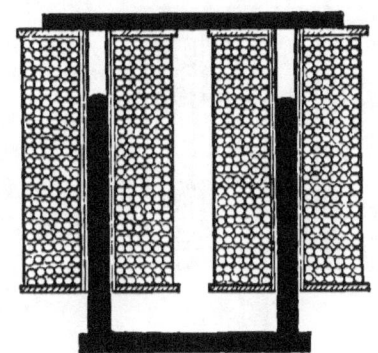

FIG. 66.—ELECTROMAGNET OF BRUSH ARC LAMP.

a fairly complete magnetic circuit, together with a comparatively long travel of the plunger and coil. It is a fair compromise between the two ways of working. The pull is not, however, in any of these forms, equal all along the whole range of travel; it increases as the magnetic circuit becomes more complete.

There are several other intermediate forms. For example, one inventor, Gaiser, employs a horseshoe electromagnet, the cores of which protrude a good distance beyond the coils, and for an armature he employs a piece of sheet iron, bent round so as to make at its ends two tubes, which inclose the poles, and are drawn down over them. Contrast with this design one of much earlier date by an engineer, Roloff, who made his electromagnets with iron cores not standing out, but sunk below the level of the ends of the coils, while the armature was furnished with little extensions that passed down into these projecting tubular ends of the coils.

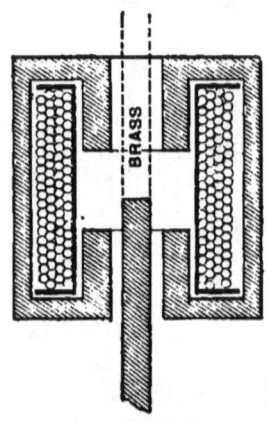

Fig. 67.—Ayrton and Perry's Tubular Iron-clad Electromagnet.

Some arc lamps have magnets of precisely that form, with a short plunger entering a tubular coil, and met half-way down by a short fixed core inside the tube.

Here (Fig. 67) is one form of tubular iron-clad electromagnet that deserves a little more attention, being the one used by Messrs. Ayrton and Perry in 1882; a coil has an iron jacket round it, and also an annular iron disc across the top, and an annular iron disc across the bottom, there being also a short internal tube of iron extending a little way down from the top, almost meeting another short internal tube of iron coming up from the bottom. The magnetic effect of the inclosed copper coil is concentrated within an extremely short space, between the ends of the internal tubes, where there is a wonderfully strong uniform field. The range of action you can alter just as you please in the construction by shortening or lengthening the internal tubes. An iron rod inserted below is drawn with great power and equality of pull over the range from one end to the other of these internal tubes.

ACTION OF MAGNETIC FIELD ON SMALL IRON SPHERE.

In dealing with the action of tubular coils upon iron cores, I showed how, when a very short core is placed in a uniform magnetic field, it is not drawn in either direction. The most extreme case is where a small sphere of soft iron is employed. Such a sphere, if placed in even the most powerful magnetic field, does not tend to move in any direction if the field is truly uniform. If the field is not uniform, then the iron sphere always tends to move from the place where the field is weak to a place where the field is stronger. A ball of bismuth or one of copper tends, on the contrary, to move from a place where the field is strong to a place where the field is weaker. This is the explanation of the actions called "dia-magnetic," which were at one time erroneously attributed to a supposed dia-magnetic polarity opposite in kind to the ordinary magnetic polarity. A simple way of stating the facts is to say that a small sphere of iron tends to move up the *slope* of a magnetic field, with a force proportional to that slope; while (in air) a sphere of bismuth or one of copper tends, with a feeble force, to move down that slope; any small piece of soft iron—a short cylinder, for example—shows the same kind of behavior as a small sphere. In some of Ayrton and Perry's coiled-ribbon ampère-meters and voltmeters, and in some of Sir William Thomson's current meters, this principle is applied.

SECTIONED COILS, WITH PLUNGER.

An important suggestion was made by Page, about 1850, when he designed a form of coil-and-plunger having a travel of indefinitely long range. The coiled tube instead of consisting merely of one coil, excited simultaneously throughout its whole length by the current, was constructed in a number of separate sections or short tubes, associated together end to end, and furnished with means for turning on the electric current into any of the sections separately. Suppose an iron core to be just entering into any section, the current is turned on in that section, and as the end of the core passes through it the current is then turned on in the section next ahead. In this way an attraction may be kept up along a tube of indefinite length. Page constructed an electric motor on this plan, which was later revived by Du Moncel, and again by Marcel Deprez in his electric "hammer."

WINDING OF TUBULAR COILS AND ELECTROMAGNETS.

The mention of this mode of winding in sections leads me to say a few final words about winding in general. All ordinary coils, whether tubular or provided with fixed cores, are wound in layers of alternate right-handed and left-handed spirals. In a preceding lecture I mentioned the mistaken notion, now disproved, that there is any gain in making all the spirals right-handed or all left-handed. For one particular case there is an advantage in winding a coil in sections; that is to say, in placing partitions or *cloisons* at intervals along the bobbin, and winding the wire so as to fill up each of the successive spaces between the partitions before passing on from one space to the next. The case in which this construction is advantageous is the unusual case of coils that are to be used with currents supplied at very high potentials. For when currents are supplied at very high potentials there is a very great tension exerted on the insulating material, tending to pierce it with a spark. By winding a coil in *cloisons,* however, there is never so great a difference of potential between the windings on two adjacent layers as there would be if the layers were wound from end to end of the whole length of coil. Consequently, there is never so great a tension on the insulating material between the layers, and a coil so wound is less likely to be injured by the occurrence of a spark.

Another variety of winding has been suggested, namely, to employ in the coils a wire of graduated thickness. It has been shown by Sir William Thomson to be advantageous in the construction of coils of galvanometers to use for the inner coils of small diameter a thin wire; then, as the diameter of the windings increases, a thicker wire; the thickest wire being used on the outermost layers; the gauge being thus proportioned to the diameter of the windings. But it by no means follows that the plan of using *graded wire,* which is satisfactory for galvanometer coils, is necessarily good for electromagnets. In designing electromagnets it is necessary to consider the means of getting rid of heat; and it is obvious that the outer layers are those which are in the most favorable position for getting rid of this heat. Experience shows that the under layers of coils of electromagnets always attain a higher temperature than those at the surface. If, therefore, the inner layers were to be wound with finer wire, offering higher resistance, and generating more heat than the outer layers, this tendency to overheating would be still more accentuated. Indeed, it would seem wise rather to reverse the galvanometer plan, and wind electromagnets with wires that are stouter on the inner layers and finer on the outer layers.

Yet another mode of winding is to employ several wires united in parallel, a separate wire being used for each layer, their anterior extremities being all soldered together at one end of the coil, and their posterior extremities being all soldered together at the other. Magnetically, this mode of winding presents not the slightest advantage over winding with a single stout wire of equivalent section. But it has lately been discovered that this mode of winding with *multiple wire* possesses one incidental advantage, namely that its use diminishes the tendency to sparking which occurs at break of circuit.

EXTENSION OF RANGE BY OBLIQUE APPROACH.

I now pass to the means which have been suggested for extending the range of motion, or of modifying its amount at different parts of the range, so as to equalize the very unequable pull. There are several such devices, some electrical, others purely mechanical, others electro-mechanical. First, there is an electrical method. André proposed that, as soon as the armature has begun to move nearer, and comes to the place where it is attracted more strongly, it is automatically to make a contact, which will shunt off part of the current and make the magnetism less powerful. Burnett proposed another means; a number of separate electromagnets acting on one armature, but as the latter approached these electromagnets were one after the other cut out of the circuit. I need not say the advantages of that method are very hypothetical. Then there is another method which has been used many times with very great success, the method of allowing the motion of the armature to occur obliquely, it being mechanically constrained so as to move past, instead of toward the pole. When the armature is pulled thus obliquely, the pull will be distributed over a definite wider range. Here is a little motor made on that very plan. A number of pieces of iron set on the periphery of a wheel are successively attracted up sideways. An automatic device breaks the circuit as every piece of iron comes near, just at the moment when it gets over the poles, and the current being cut off, it flies on beyond and another piece comes up, is also attracted in the same way, and then allowed to pass. A large number of toy motors have been made from time to time on this plan. I believe Wheatstone was the first to devise the method of oblique approach about the year 1841. He made many little electromagnetic motors, the armatures of which were in some cases solid rims of iron arranged as a sort of wheel, with two or more zigzag internal teeth, offering oblique surfaces to the attraction of an electromagnet. Such little motors are often now used for spinning Geissler's vacuum tubes. In these motors the iron rim is fixed and the electromagnet rotates. The pole of the electromagnet finds itself a certain distance away from the iron ring; it tries to get nearer. The only way it can get nearer is by swinging round, and so it gradually approaches, and as it approaches the place where it is nearest to the internal projection of the rim the current is cut off, and it swings further. This mode may be likened to a cam in a mechanical movement. It is, in fact, nothing else than an *electromagnetic cam*. There are other devices too, which are more like electromagnetic *linkage*. If you curve the poles or shape them out, you may obtain actions which are like that of a wedge on an inclined plane. There is an electromagnet in one of Paterson and Cooper's arc lamps wherein the pole-piece, coming out below the magnet, has a very peculiar shape, and the armature is so pivoted with respect to the magnet, that as the armature approaches the core as a whole its surface recedes from that of the pole-piece, the effect being that the pull is equalized over a considerable range of motion. There is a somewhat similar device in De Puydt's pattern of arc lamp.

Here is another device for oblique approach, made by Froment. In the gap in the circuit of the magnet a sort of iron wedge is put in, which is not attracted squarely to either face, but comes in laterally between guides. Another of Froment's equalizers, or distributors, consists of a parallel motion attachment for the armature, so that oblique approach may take place, without actual contact. Here (Fig. 68) is another mechanical method of equalizing devised by Froment, and used by Le Roux. You know the Stanhope lever, the object of which is to transform a weak force along a considerable range into a powerful force of short range. Here we use it backward. The armature itself, which

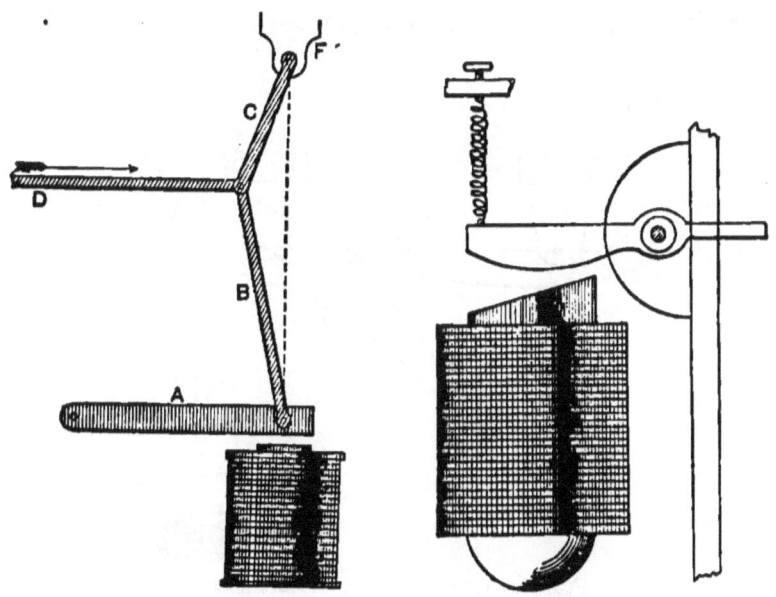

FIG. 68.—FROMENT'S EQUALIZER WITH STANHOPE LEVER.

FIG. 69.—DAVY'S MODE OF CONTROLLING ARMATURE BY SPRING.

is attracted with a powerful force of short range, is attached to the lower end of the Stanhope lever, and the arm attached to the knee of the lever will deliver a distributed force over quite a different range. One way, not of equalizing the actual motion over the range, but of counterbalancing the variable attractive force, is to employ a spring instead of gravity to control the armature. So far back as 1838, Edward Davy, in one of his telegraphic patents, described the use of a spring (Fig. 69) to hold back the armature. Davy preceded Morse in the use of a spring to pull back the armature. There is a way of making a spring act against an armature more stiffly as the pull gets greater. In this method there is a spring with various set screws set up against

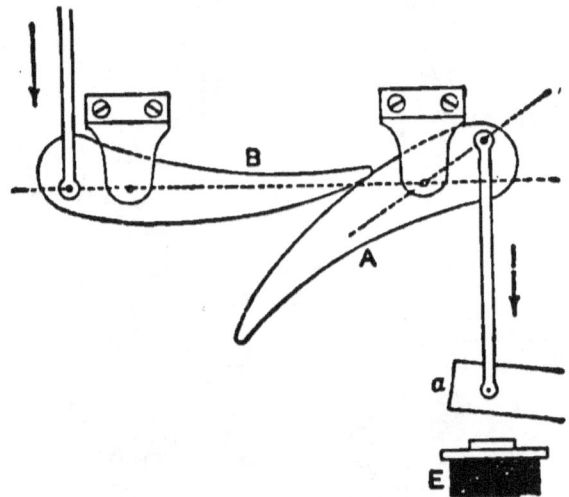

Fig. 70.—Robert Houdin's Equalizer.

it, and which come into action at different ranges, so as to alter the stiffness of the spring, making it virtually stiffer as the armature approaches the poles. Yet another method is to employ, as the famous conjurer Robert Houdin did, a rocking lever. Fig. 70 depicts one of Robert Houdin's equalizers. The pull of the electro-magnet on the armature acts on a curved lever which works against a second one, the point of application of force between the one and the other altering with their position. When the armature is far away from the pole, the leverage of the first lever on the second lever is great. When the armature gets near, the leverage of the first lever on the second is comparatively small. This employment of the *rocking lever* was adopted from Houdin by Duboscq, and put into the Duboscq arc lamp, where the regulating mechanism at the bottom of the lamp contains a rocking lever. Here upon the lecture table is a Duboscq arc lamp. In this pattern (Fig. 71),

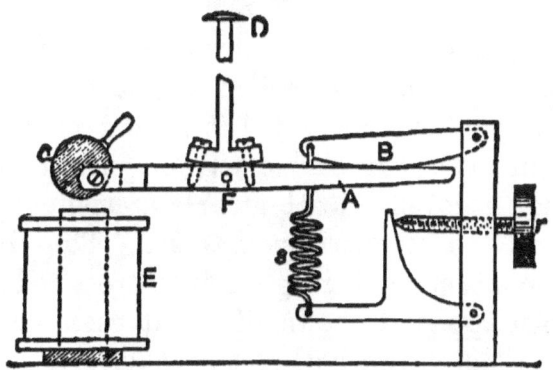

FIG. 71.—MECHANISM OF DUBOSCQ'S ARC LAMP.

one lever, *B*, which is curved, plays against another, *A*, which is straight. A similar mechanism is used for equalizing the action in the Serrin arc lamp, where one of the springs that holds up the jointed parallelogram frame is applied at the end of a rocking lever to equalize the pull of the regulating electromagnet. In this lamp there is also introduced the principle of oblique approach; for the armature of the electromagnet is not allowed to travel straight toward the poles of the magnet, but is pulled up obliquely past it.

Another device for equalizing the pull was used by Wheatstone in the step-by-step telegraph in 1840. A hole is pierced in the armature, and the end of the core is formed into a projecting cone, which passes through the aperture of the armature, thereby securing a more equable force and a longer range. The same device has reappeared in recent years in the form of electromagnet used in the Thomson-Houston arc lamp, and in the automatic regulator of the same firm.

POLARIZED MECHANISM: USES OF PERMANENT MAGNETS.

We must now turn our attention to one class of electromagnetic mechanism which ought to be carefully distinguished from the rest. It is that class in which, in addition to the ordinary electromagnet, a permanent magnet is employed. Such an arrangement is generally referred to as a *polarized mechanism*. The objects for which the permanent magnet is introduced into the mechanism appear to be in different cases quite different. I am not sure whether this is clearly recognized, or whether a clear distinction has even been drawn between three entirely separate purposes in the use of a permanent magnet in combination with an electromagnet. The first purpose is to secure unidirectionality of motion; the second is to increase the rapidity of action and of sensitiveness to small currents; the third to augment the mechanical action of the current.

(*a*.) *Unidirectionality of Motion.*—In an ordinary electromagnet it does not matter which way the current circulates; no matter whether the pole is north or south, the armature is pulled, and on reversing the current the armature is also pulled. There is a rather curious old experiment which Sturgeon and Henry showed, that if you have an electromagnet with a big weight hanging on it, and you suddenly reverse the current, you reverse the magnetism, but it still holds the weight up; it does not drop. It has not time to drop before the magnet is charged up again with magnetic lines the other way on. Whichever way the magnetism traverses the ordinary soft iron electromagnet, the armature is pulled. But if the armature is itself a permanent magnet of steel, it will be pulled when the poles are of one sort, and pushed when the poles are reversed—that is to say, by employing *a polarized armature* you can secure unidirectionality of motion in correspondence with the current. One immediate application of this fact for telegraphic purposes is that of duplex telegraphy. You can send two messages at the same time and in the same direction to two different sets of instruments, one set having ordinary electromagnets, with a spring behind the armature of soft iron, which will act simply independently of the direction of the current, depending only on its strength and duration; and another set having electromagnets with polarized armatures, which will be affected not by the strength of the current, but by the direction of it. Accordingly, two completely different sets of messages may be sent through that line in the same direction at the same time.

Another mode of constructing a polarized device is to attach the cores of the electromagnet to a steel magnet, which imparts to them an initial magnetization. Such initially magnetized electromagnets were used by Brett in 1848 and by Hjörth in 1850. A patent for a similar device was applied for in 1870 by Sir William Thomson and refused by the Patent Office. In 1871 S. A. Varley patented an electromagnet having a core of steel wires united at their ends.

Wheatstone used a polarized apparatus consisting of an electromagnet acting on a magnetized needle. He patented, in fact, in 1845, the use of a needle permanently magnetized to be attracted one way or the other between the poles of an electromagnet. Sturgeon had described the very same device in the *Annals of Electricity* in 1840. Gloesner claims to have invented the substitution of permanent magnets for mere armatures in 1842. In using polarized apparatus it is necessary to work, not with a simple current that is turned off and on, but with reversed currents. Sending a current one way will make the moving part move in one direction; reversing the current makes it go over to the other side. The mechanism of that particular kind of electric bell that is used with magneto-electric calling apparatus furnishes an excellent example of a polarized construction. With these bells no battery is used; but there is a little alternate current dynamo, worked by a crank. The alternate currents cause the pivoted armature in the bell to oscillate to right and left alternately, and so throw the little hammer to and fro between the two bells.

(*b.*) *Rapidity and Sensitiveness of Action.*—For relay work polarized relays are often employed, and have been for many years. Here on the table is one of the post-office pattern of standard relay, having a steel magnet to give magnetism permanently to a little tongue or armature which moves between the poles of an electromagnet that does the work of receiving the signals. In this particular case the tongue of the polarized relay works between two stops, and the range of motion is made very small in order that the apparatus may respond to very small currents. At first sight it is not very apparent why putting a permanent magnet into a thing should make it any more sensitive. Why should permanent magnetism secure rapidity of working? Without knowing anything more, inventors will tell you that the presence of a permanent magnet increases the rapidity with which it will work. You might suppose that permanent magnetism is something to be avoided in the cores of your working electromagnets, otherwise the armatures would remain stuck to the poles when once they had been attracted up. Residual magnetism would, indeed, hinder the working unless you have so arranged matters that it shall be actually helpful to you. Now for many years it was supposed that permanent magnetism in the electromagnet was anything but a source of help. It was supposed to be an unmitigated nuisance, to be got rid of by all available means, until, in 1855, Hughes showed us how very advantageous it was to have permanent magnetism in the cores of the electromagnet. Here (Fig. 51), is the drawing of Hughes' magnet to which I referred in Lecture III. A compound permanent magnet of horseshoe shape is provided with coils on its pole-pieces, and there is a short armature on the top attached to a pivoted lever and a counteracting spring. The function of this arrangement is as follows: That spring is so set as to tend to detach the armature, but the permanent magnet has just enough magnetism to hold the armature on. You can, by screwing up a little screw behind the spring, adjust these two contending forces, so that they are in the nicest possible balance; the armature held on by the magnetism, and the spring just not able to pull it off. If, now, when these two actions are so nearly balanced you send an electric current round the coils, if the electric current goes one way round it just weakens the magnetism enough for the spring to gain the victory, and up goes the armature. This apparatus then acts by letting the armature off when the balance is upset by the electric current; and it is capable of responding to extremely small currents. Of course, the armature has to be put on again mechanically, and in Hughes' type-writing telegraph instruments it is put on mechanically between each signal and the next following one. The arrangement constitutes a distinctive piece of electromagnetic mechanism.

(c.) *Augmenting Mechanical Action of Current.*—The third purpose of a permanent magnet, to secure a greater mechanical action of the varying current, is closely bound up with the preceding purpose of securing sensitiveness of action. It is for this purpose that it is used in telephone receivers; it increases the mechanical action of the current, and therefore makes the receiver more sensitive. For a long time this was not at all clear to me, indeed I made experiments to see how far it was due to any variation in the magnetic permeability of iron at different stages of magnetization, for I found that this had something to do with it, but I was quite sure it was not all. Prof. George Forbes gave me the clue to the true explanation; it lies in the law of traction with which you are now familiar, that the pull between a magnet and its armature is proportional to the square of the number of magnetic lines that come into action. If we take **N**, the number of magnetic lines that are acting through a given area, then to the square of that the pull will be proportional. If we have a certain number of lines, **N**, coming permanently to the armature, the pull is proportional to $\mathbf{N}^2$. Suppose the magnetism now to be altered—say made a little more; and the increment be called $d\mathbf{N}$; so that the whole number is now $\mathbf{N}+d\mathbf{N}$. The pull will now be proportional to the square of that quantity. It is evident that the motion will be proportional to the difference between the former pull and the latter pull. So we will write out the square of $\mathbf{N}+d\mathbf{N}$ and the square of **N** and take the difference.

Increased pull, proportional to	$\mathbf{N}^2 + 2\,\mathbf{N}d\mathbf{N} + d\mathbf{N}^2;$
Initial pull, proportional to	$\mathbf{N}^2$
Subtracting; difference is	$2\,\mathbf{N}d\mathbf{N} + d\mathbf{N}^2.$

We may neglect the last term, as it is small compared with the other. So we have, finally, that the change of pull is proportional to $2\,\mathbf{N}d\mathbf{N}$. The alteration of pull between the initial magnetism and the initial magnetism with the additional magnetism we have given to it turns out to be proportional not simply to the change of magnetism, but also to the initial number **N**, that goes through it to begin with. The more powerful the pull to begin with, the greater is the change of pull when you produce a small change in the number of magnetic lines. That is why you have this greater sensitiveness of action when using Hughes' electromagnets, and greater mechanical effect as the result of applying permanent magnetism to the electromagnets of telephone receivers.

ELECTROMAGNETIC MECHANISM.

There are some other kinds of electromagnetic mechanism to which I must briefly invite your attention as forming an important part of this great subject. Of one of these the mention of permanent magnets reminds me.

MOVING COIL IN PERMANENT MAGNETIC FIELD.

A coil traversed by an electric current experiences mechanical forces if it lies in a magnetic field, the force being proportional to the intensity of the field. Of this principle the mechanism of Sir Wm. Thomson's siphon recorder is a well-known example. Also those galvanometers which have for their essential part a movable coil suspended between the poles of a permanent magnet, of which the earliest example is that of Robertson ("Encyclopædia Britannica," ed. viii., 1855), and of which Maxwell's suggestion, afterward realized by d'Arsonval, is the most modern. Siemens has constructed a relay on a similar plan.

MAGNETIC ADHERENCE.

There are a few curious pieces of apparatus devised for increasing adherence electromagnetically between two things. Here is an old device of Nicklès, who thought he would make a new kind of rolling gear. Whether it was a railway wheel on a line, or whether it

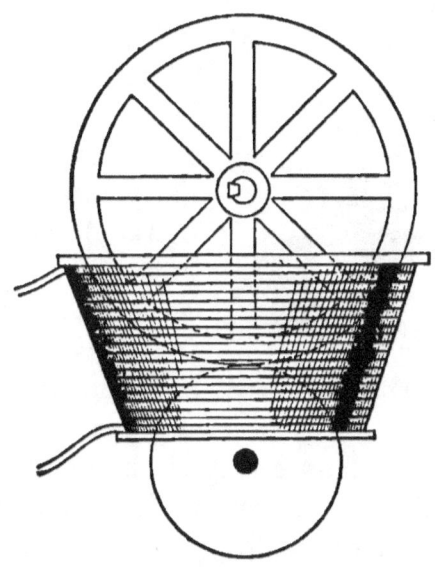

FIG. 72.—NICKLÈS' MAGNETIC FRICTION GEAR.

was going to be an ordinary wheel gearing, communication of motion was to be made from one wheel to another, not by cogs or by the mere adherence of ordinary friction, but by magnetic adherence. In Fig. 72 there are shown two iron wheels rolling on one another, with a sort of electromagnetic jacket around them, consisting of an electric current circulating in a coil, and causing them to attract one another and stick together with magnetic adherence. In Nicklès' little book on the subject there are a great number of devices of this kind described, including a magnetic brake for braking railway wagons, engines, and carriages, applying electromagnets either to the wheels or else to the line, to stop the motion whenever desired. The notion of using an electromagnetic brake has been revived quite recently in a much better form by Prof. Geo. Forbes and Mr.

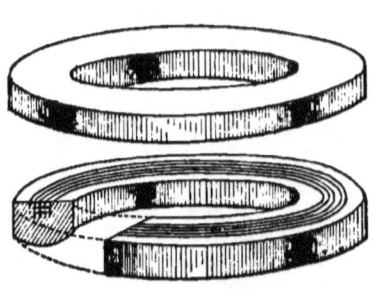

FIG. 73.—FORBES' ELECTROMAGNET.

Timmis, whose particular form of electromagnet, shown in Fig. 73, is peculiarly interesting, being a better design than any I have ever seen for securing powerful magnetic traction for a given weight of iron and copper. The magnet is a peculiar one; it is represented here as cut away to show the internal construction. There is a sort of horseshoe made of one grooved rim, the whole circle of coil being laid imbedded in the groove. The armature is a ring which is attracted down all round, so that you have an extremely compact magnetic circuit around the copper wire at every point. The magnet part is attached to the frame of the wagon or carriage, and the ring-armature is attached to the wheel or to its axis. On switching on the electric current the rim is powerfully pulled, and braked against the polar surface of the electromagnet.

Forbes' arrangement appears to be certainly the best

LECTURES ON THE ELECTROMAGNET. 273

yet thought of for putting a magnetic brake to the wheels of a railway train.

Another, but quite distinct, piece of mechanism depending on electromagnetic adherence is the magnetic *clutch* employed in Gülcher's arc lamp.

REPULSION MECHANISM.

Then there are a few pieces of mechanism which depend on repulsion. In 1850 a little device was patented by Brown and Williams, consisting, as shown in Fig.

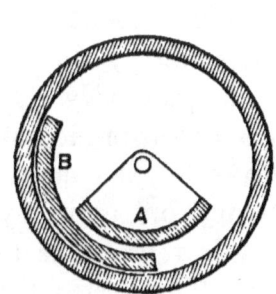

Fig. 74.—Electromagnetic Mechanism Working by Repulsion.

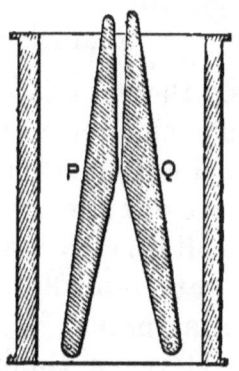

Fig. 75.—Repulsion Between Two Parallel Cores.

74, of an electromagnet which repelled part of itself. The coil is simply wound on a hollow tube, and inside the coil is a piece, *B*, of iron, bent as the segment of a cylinder to fit in, going from one end to the other. Another little iron piece, *A*, also shaped as the segment of a tube, is pivoted in the axis of the coil. When these are magnetized one tends to move away from the other, they being both of the same polarity. Of late there have been many ampère-meters and voltmeters made on this plan of producing repulsion between the parallel cores.

Here (Fig. 75) is another device of recent date, due to Maikoff and De Kabath. Two cores of iron, not quite parallel, pivoted at the bottom, pass up through a tubular coil. When both are magnetized, instead of attracting one another, they open out; they tend to set themselves along the magnetic lines through that tube. The cores, being wide open at the bottom, tend to open also at the top.

ELECTROMAGNETIC VIBRATORS.

Then there is a large class of mechanisms about which a whole chapter might be written, namely, those in which vibration is maintained electromagnetically. The armature of an electromagnet is caused to approach and recede alternately with a vibrating motion, the current being automatically cut off and turned on again by a self-acting brake. The electromagnetic vibrator is one of the cleverest things ever devised. The first vibrating electromagnetic mechanism ever made was exhibited here in this room in 1824 by its inventor, an Englishman named James Marsh. It consisted of a pendulum vibrating automatically between the poles of a permanent magnet. Later, a number of other vibrating devices were produced by Wagner, Neef, Froment, and others. Most important of all is the mechanism of the common electric trembling bell, invented by a man whose very name appears to be quite forgotten—John Mirand. How many of the millions of people who use electric bells know the name of the man who invented them? John Mirand, in the year 1850, put the electric bell practically into the same form in which it has been employed from that day to this. The vibrating hammer, the familiar push-button, the indicator or annunciator, are all of his devising, and may be seen depicted in the specification of his British patent, just as they came from his hand.

Time alone precludes me from dealing minutely with these vibrators, and particularly with the recent work of Mercadier and that of Langdon-Davies, whose researches have put a new aspect on the possibilities of harmonic telegraphy. Langdon-Davies' rate governor is the most recent and perfect form of electromagnetic vibrator.

INDICATOR MOVEMENTS.

Upon the table here are a number of patterns of electric bells, and a number also of the electro-mechanical movements or devices employed in electric bell work, some of which form admirable illustrations of the various principles that I have been laying down. Here is an iron-clad electromagnet; here a tripolar magnet; here a series of pendulum motions of various kinds; here is an example of oblique pull; here is Jensen's indicator, with lateral pull; here is Moseley's indicator, with a coil-and-plunger, iron-clad; here is a clever device in which a disc is drawn up to better the magnetic circuit. Here, again, is Thorpe's semaphore indicator, one of the neatest little pieces of apparatus, with a single central core surrounded by a coil, while a little strip of iron coming round from behind serves to complete the circuit all save a little gap. Over the gap stands that which is to be attracted, a flat disc of iron, which, when it is attracted, unlatches another disc of brass which forthwith falls down. It is an extremely effective, very sensitive, and very inexpensive form of annunciator. The next two are pieces of polarized mechanism having a motion directed to one side or the other, according to the direction of the current. From the backboard projects a small straight electromagnet. Over it is pivoted a small arched steel magnet, permanently magnetized, to which is attached a small signal lever bearing a red disc. If there is a current flowing one way then the magnet that straddles over the pole of the electromagnet will be drawn over in one direction. If I now reverse the current the electromagnet attracts the other pole of the curved magnet. Hence this mechanism allows of an electrical replacement without compelling the attendant to walk up to the indicator board. The polarized apparatus for indicators has this advantage, that you can have electrical as distinguished from mechanical replacement.

THE STUDY OF ELECTROMAGNETIC MECHANISM.

The rapid survey of electromagnetic mechanisms in general has necessarily been very hurried and imperfect. The study of it is just as important to the electrical engineer as is the study of mechanical mechanism to the mechanical engineer. Incomplete as is the present treatment of the subject, it may sufficiently indicate to other workers useful lines of progress, and so fitly be appended to these lectures on the electromagnet. In a very few years we may expect the introduction into all large engineering shops of *electromagnetic tools*. On a small scale, for driving dental appliances, electromagnetic engines have long been used. Large machine tools, electromagnetically worked, have already begun to make their appearance. Some such were shown at the Crystal Palace, in 1881, by Mr. Latimer Clark, and more recently Mr. Rowan, of Glasgow, has devised a number of more developed forms of electromagnetic tools.

SUPPRESSION OF SPARKING.

It now remains for me to speak briefly of the suppression of sparks. There are some half-dozen different ways of trying to get rid of the sparking that occurs in the breaking of an electric circuit whenever there are electromagnets in that circuit. Many attempts have been made to try and get rid of this evil. For instance, one inventor employs an air blast to blow out the spark just at the moment it occurs. Another causes the spark to occur under a liquid. Another wipes it out with a brush of asbestos cloth that comes immediately behind the wire and rubs out the spark. Another puts on a condenser to try and store up the energy. Another tries to put on a long thin wire or a high resistance of liquid, or something of that kind, to provide an alternate path for the spark, instead of jumping across the air and burning the contacts. There exist some half-score, at any rate, of that kind of device. But there are devices that I have thought it worth while to examine and experiment upon, because they depend merely upon the mode of construction adopted in the building of the electromagnet, and they have each their own qualities. I have here five straight electromagnets, all wound on bobbins the same size, for which we shall use the same iron core and the same current for all. They are all made, not only with bobbins of the same size, but their coils consist as nearly as possible of the same weight of wire. The first one is wound in the ordinary way; the second one has a sheath of copper wound round the interior of the bobbin before any wire is put on. This was a device, I believe, of the late Mr. C. F. Varley, and is also used in the field magnets of Brush dynamos. The function of the copper sheath is to allow induced current to occur, which will retard the fall of magnetism, and damp down the tendency to spark. The third one is an attempt to carry out that principle still further. This is due to an American of the name of Paine, and has been revived of late years by Dr. Aron, of Berlin. After winding each layer of the coil, a sheath of metal foil is interposed so as to kill the induction from layer to layer. The fourth one is the best device hitherto used, namely, that of differential winding, having two coils connected so that the current goes opposite ways. When equal currents flow in both circuits there is no magnetism. If you break the circuit of either of the two wires the core at once becomes magnetized. You get magnetism on breaking, you destroy magnetism on making the circuit; it is just the inverse case to that of the ordinary electromagnet. There the spark occurs when magnetism disappears, but here, since the magnetism disappears when you make the circuit, you do not get any spark at make, because the circuit is already made. You do not get any at break, because at break there is no magnetism. The fifth and last of these electromagnets is wound according to a plan devised by Mr. Langdon-Davies, to which I alluded in the middle of this lecture, the bobbin being wound with a number of separate coils in parallel with one another, each layer being a separate wire, the separate ends of all the layers being finally joined up. In this case there are 15 separate circuits; the time-constants of them are different, because, owing to the fact that these coils are of different diameters, the coefficient of self-induction of the outer layers is rather less, and their resistance, because of the larger size, rather greater than those of the inner layers. The result is that instead of the extra current running out all at the same time, it runs out at different times for these 15 coils. The total electromotive force of self-induction never rises so high and it is unable to jump a large air-gap, or give the same bright spark as the ordinary electromagnet would give. We will now experiment with these coils. The differential winding gives absolutely no spark at all, and second in merit comes No. 5, with the multiple wire winding. Third in merit comes the coil with intervening layers of foil. The fourth is that with copper sheath. Last of all, the electromagnet with ordinary winding.

CONCLUSION.

Now let me conclude by returning to my starting-point—the invention of the electromagnet by William Sturgeon. Naturally you would be glad to see the counterfeit presentment of the features of so remarkable a man, of one so worthy to be remembered among distinguished electricians and great inventors. Your disappointment cannot be greater than mine when I tell you that all my efforts to procure a portrait of the deceased inventor have been unavailing. Only this I have been able to learn as the result of numerous inquiries; that an oil-painting of him existed a few years ago in the possession of his only daughter, then resident in Manchester, whose address is now, unfortunately, unknown. But if his face must remain unknown to us, we shall none the less proudly concur in honoring the memory of one whose presence once honored this hall wherein we are met, and whose work has won for him an imperishable name.

INDEX.

AIR-GAP, effect of, in magnetic circuit, 221
    effect of, on magnetic reluctance, 117, 119, 144
André, equalizing the pull of a magnet, 258
Ampère, researches of, 16
Ampère turns, calculation of, 166
Arago, researches of, 16
Arc lamp mechanism, 54
    Brockie-Pell, 250
    Brush, 253
    De Puydt, 260
    Duboscq, 263
    Gaiffe, 248
    Gülcher, 273
    Paterson and Cooper, 250, 260
    Pilsen, 247
    Serrin, 263
    Thomson-Houston, 264
    Weston, 253
Armature, effect of, on permanent magnets, 200
    effect of shape, 80
    length and cross-section of, 196
    position and form of, 197
    pulled obliquely, 259
    round vs. flat, 197
Aron, sheath for magnet coils, 278
Ayrton, distribution of free magnetism, 109
    magnetic shunts, 13
Ayrton and Perry's coiled ribbon voltmeters, 255
    tubular electromagnet, 254

BAR electromagnet, 49
    Barlow, magnetism of long bars, 151
Barlow's wheel, 16
Battery grouping for quickest action, 215
    resistance for best effect, 78, 185
    used by Sturgeon, 18
Bell (A. G.), iron-clad electromagnet, 135
Bernoulli's rule for traction, 98
Bidwell, electromagnetic pop-gun, 206
    measurement of permeability, 68
Bosanquet, investigations of, 90
    magneto-motive force, 12
    measurement of permeability, 59, 63
Brett, polarized magnets, 266
Brisson, method of winding, 184
Brockie-Pell, differential coil-and-plunger, 250
Brown and Williams, repulsion mechanism, 273
Bruger, coils and plungers, 244
Burnett, equalizing the pull of a magnet, 259

CAMACHO'S electromagnet, 202
    Cance's electromagnet, 202
Cast iron, magnetization of, 56
Clark, electromagnetic tools, 277
Coil-and-plunger coil, 251
    diagram of force and work of, 235
    differential, 250

282 INDEX.

Coil-and-plunger electromagnet, 50, 232, 238, 242, 244
  modifications of, 250
Coil moved in permanent magnetic field, 270
Coils, effect of position, 192
  effect of size, 191
  how connected for quickest action, 213
Coned plungers, effect of, 246
Cook's experiments, 38
Cores, effect of shape, 80
  determination of length, 154
  effect of shape of section, 193
  hollow versus solid, 78
  lamination of, 207, 213
  of different thicknesses, 243
  of irregular shapes, 248
  proper length of, 94, 118
  square versus round, 78
  tubular, 158
Coulomb, law of inverse squares, 111
  two magnetic fluids, 9
Cowper, lamination of cores, 207
  range of action, 225
Cumming, magnetic conductivity, 10
  galvanometer, 16
Curves of hysteresis, 75
  of magnetization and permeability, 71

D'ARSONVAL, galvanometer, 270
  Davy, mode of controlling armature, 261
Davy, researches of, 16
De La Rive, floating battery and coil, 16
  magnetic circuit, 10
Deprez, electric hammer, 256
Diacritical point of magnetization, 74
Diamagnetic action, 255
Dove, magnetic circuit, 10
Dub, best position of coils, 192
  cores of different thicknesses, 243
  distance between poles, 194

Dub, flat vs. pointed poles, 125, 127
  magnetic circuit, 10
  magnetism of long bars, 152
  polar extensions of core, 126
  thickness of armatures, 158
Du Moncel, best position of coils, 192
  club-footed electromagnet, 189
  distance between poles, 194
  effect of polar projections, 198
  effect of position of armature, 151
  electric motor, 256
  electromagnetic pop-gun, 205
  experiments with pole-pieces, 132
  length of armatures, 158
  on armatures, 197
  tubular cores, 159

ELECTRIC bells, 275
  invented by Mirand, 274
Electric indicators, 275
  motors, not practicable, 225
Electromagnet, Ayrton and Perry's, 254
  bar, 49, 188
  Camacho's, 202
  Cance's, 202
  club-footed, 189
  coil-and-plunger, 50, 222
  coils, resistance of, 78
  design of, for various uses, 9
  Du Moncel's, 189
  Fabre's, 135
  Faulkner's, 135
  first publicly described, 7, 17
  for rapid working, 195
  Gaiser's, 253
  Guillemin's, 135
  Henry's, 27
  Hjörth's, 224
  horseshoe, 49, 188
  Hughes', 195, 267
  in Bell's telephone, 135
  invented in 1825, 16
  iron-clad, 50, 78, 133, 135, 188
  Jensen's, 191

INDEX. 283

Electromagnet, Joule's, 39, 46
  law of, 8
  long vs. short limbs, 154
  of Brush arc lamp, 207
  Radford's, 46
  Roberts', 46
  Roloff's, 254
  Romershausen's, 135
  Ruhmkorff's, 204
  Smith's, 253
  Stevens and Hardy, 252
  Sturgeon's, 18
  Varley's, 188, 202
  Wagener's, 204
  without iron, 251
Electromagnetic clutch, 273
  engines, 223
  inertia, 187, 208
  mechanism, 222, 270
  pop-gun, 205
  repulsion, 208
  tools, 277
  vibrators, 274
Electromagnets, diminutive, 100
  fallacies and facts about, 77
  for alternating currents, 206
  for arc lamp (see arc lamp mechanism)
  for lifting, 52
  for maximum range of attraction, 203
  for maximum traction, 202
  for minimum weight, 203
  formulæ for, 74
  for quickest action, 209
  for traction, 41, 52
  heating of, 76
  in telegraph apparatus, 207, 221
  saturation of, 44
  specifications of, 185
  to produce rapid vibrations, 53
  with iron between the windings, 212
  with long versus short limbs, 79, 171, 220

Elphinstone, Lord, application of magnetic circuit in dynamo design, 12
Equalizing the pull of a magnet, 258
Ewing, curves of magnetization, 57
  hysteresis, 75
  maximum magnetization, 72
  measurement of permeability, 59, 63
  on effect of joints, 155

FABRE, iron-clad electromagnet, 135
Faraday, lines of force, 11
  rotation of permanent magnet, 16
Faulkner, iron-clad electromagnet, 135
Forbes, electromagnetic brake, 272
  formulæ for estimation of leakage, 144
  magnetic leakage, 13
  polarized apparatus, 269
Frölich, law of the electromagnet, 73
Froment s equalizer, 260
  vibrating mechanism, 274

GAISER'S electromagnet, 253
  Galvanometer coils, 270
Gauss, magnetic measurements, 113
Gloesner, polarized magnets, 266
Grove, range of action, 225
Guillemin, iron-clad electromagnet, 135

HÄCKER'S rule for traction, 98
  Hankel, magnetism of long bars, 152
Hankel, working of coil-and-plunger, 242
Heating of magnet coils, 96, 98, 173, 174, 175, 176, 183, 204, 207, 208, 257
Heaviside, magnetic reluctance, 82
Helmholtz, law regarding interrupted currents, 8

INDEX.

Henry's first experiments, 27
Hjörth's electromagnet, 224
    polarized magnets, 206
Hopkinson, curves of magnetization, 65
    design of dynamos, 13
    maximum magnetization, 72
    measurement of permeability, 59, 63
Horseshoe electromagnet, 49
Houdin's equalizer, 262
Hughes, distance between poles, 194
    magnetic balance, 59
    polarized magnet, 207
    printing telegraph magnets, 194, 196
Hunt, range of action of electromagnets, 224
Hysteresis, 75
    viscous, 77

IRON-CLAD electromagnet, 50, 78, 133, 135
    range of action, 249
Iron, magnetic qualities affected by hammering, rolling, etc., 77
    maximum magnetization of, 92
    permeability of, 92
    permeability of, compared with air, 85, 118
    the magnetic properties of, 54, 56

JENSEN'S electromagnet, 191
    indicator, 275
Joints, effect of, on magnetic reluctance, 155
Joule, experiment with Sturgeon's magnet, 20
    lamination of cores, 207
    law of mutual attraction, 40
    law of traction, 100
    length of electromagnet, 94
    magnetic saturation, 56
    maximum magnetization, 72

Joule, maximum power of an electromagnet, 11
    range of action, 225
    researches, 39, 81
    results of traction experiments, 91
    tubular cores, 158

KAPP, design of dynamos, 13
    maximum magnetization, 72
Keeper, effect of position on tractive power, 78
    effect of removing suddenly, 78, 202
Kirchhoff, measurement of permeability, 59
Krizik, coned and cylindrical plungers, 247

LANGDON-DAVIES' rate governor, 275
    suppression of sparking, 279
Laplace, two magnetic fluids, 9
Law of inverse squares, 13, 78, 110, 112, 226, 251
    a point law, 111
    apparatus to illustrate, 113, 115
    defined, 111
Law of the electromagnet, 73
Law of the magnetic circuit, applied to traction, 87
    as stated by Maxwell, 88
    explanation of symbols, 82
Law of Helmholtz, 209
Law of traction, 71, 100, 101, 102
    verified, 90
Leakage of magnetic lines, 85, 1 8, 110, 112, 129
Leakage reluctances, 148
Lemont, law of the electromagnet, 73
Lenz, magnetism of long bars, 151
Leupold, winding for range of action, 248
Lines of force, 11, 55
Lyttle's patent for winding, 184

INDEX.

MAGNETIC adherence, 271
    balance of Prof. Hughes, 59
  brake, 272
  centre of gravity, 112, 113
  circuit, 10, 11, 12, 13, 47
    application of, in dynamo design, 12, 13
    for greatest traction, 97
    formulæ for, 86, 87, 101, 102
    tendency to become more compact, 123, 204
    various parts of, 49
  conductivity, 10, 11, 83
  field, action of, on small iron sphere, 255
  flux, calculation of, 83, 164
  gear, 271
  insulation, 84
  leakage, 13
    calculation of, 122, 161
    calculation of coefficient, 168
    coefficient of, "v," 145
    due to air-gaps, 120, 144
    estimation of, 144, 150, 169
    measurement of, 137 [193
    proportional to the surface,
    relation of, to pull, 139
  memory, 172, 221
  moments, 13, 87, 158
  output of electromagnets, 185
  permeability, 11, 54, 83
  polarity, rule for determining, 51
  pole of the earth, 113
  reluctance, calculation of, 83, 93, 165
    of divided iron ring, 117
    of iron ring, 117
    of waste and stray field, formulæ for, 146, 150
  resistance, 12, 82
  saturation, 47, 56, 58
  shunts, 13
Magnetism, free, 9
  of long iron bars, 151
Magnetization and magnetic traction, tabular data, 89

Magnetization, calculation of, 161, 164
  defined, 87
  internal, 9, 54
  internal distribution of, 68, 78, 138
  of different materials, 57
  surface, 9, 13, 48
Magnetometer, 114
Magneto-motive force, 11, 12, 81
  calculation of, 82, 83
Maikoff and De Kabath, repulsion mechanism, 274
Marsh, first vibrating mechanism, 274
  vibrating pendulum, 16
Maxwell, galvanometer, 270
  law of the magnetic circuit stated, 88
  law regarding circulation of alternating currents, 8
  magnetic conductivity, 11
Mirand, inventor of the electric bell, 274
Mitis metal, magnetization of, 72
Moll's experiments, 22, 34
Moseley's indicator, 275
Müller, law of the electromagnet, 73
  magnetism of long bars, 152
  measurement of permeability, 58

NEEF, vibrating mechanism, 274
  Newton's signet ring loadstone, 100
Nicklès, classification of magnets, 188
  distance between limbs of horseshoe magnets, 158
  distance between poles, 194
  length of armatures, 158
  magnetic brake, 272
  magnetic gear, 271
  traction affected by extent of polar surface, 104
  tubular cores, 158

OBLIQUE approach, 258, 263
  Oersted's discovery, 16
Ohm's law, 8, 12, 26, 81, 209

PAGE, electric motor, 256
  sectioned coils, 256
  electromagnetic engine, 223
Paine, sheath for magnets, 278
Permanent magnets contrasted with electromagnets, 199
  uses of, 264
Permeability, calculation of, 103
  methods of measuring, 58
Permeameter, 70
Permeance, of telegraph instrument magnets, 150
Perry, magnetic shunts, 13
Pfaff, tubular cores, 158
Plungers, coned vs. cylindrical, 247
  of iron and steel, 244
Point poles, 114, 115
  action of single coil on, 230
Poisson, two magnetic fluids, 9
Polar distribution of magnetic lines, 137
  region, defined, 112, 113
Polarized apparatus for indicators, 276
  mechanism, 264
Pole-pieces, convex versus flat, 79, 104
  Dub's experiments with, 126
  Du Moncel's experiments with, 132
  effect of position on tractive power, 79
  effect on lifting power, 78
  on horseshoe magnets, 198
Poles, effect of distance between, 194
  flat vs. pointed, 125, 127
Preece, self-induction in relays, 218
  winding of coils, 184

Radford's electromagnet, 46
  Range of action of electromagnets, 224, 225, 248
Rate governor, 275
Reluctance, 12, 82
Repulsion mechanism, 273
Residual magnetism, 67

Resistance of electromagnet and battery, 185
  of insulated wire, rule for, 176
Ritchie, magnetic circuit, 10
  steel magnets, 172
Roberts' electromagnet, 46
Robertson, galvanometer, 270
Roloff's electromagnet, 254
Romershausen, iron-clad electromagnet, 135
Rowan, electromagnetic tools, 277
Rowland, analogy of magnetic and electric circuits, 12
  first statement of the law of the magnetic circuit, 81
  magnetic permeability, 11
  maximum magnetization, 72
  measurement of permeability, 59, 63
Ruhmkorff's electromagnet, 204

SATURATION, curve of, 153
  distribution of, 138
  effect of, on permeability, 118
Schweigger's multiplier, 16, 28
Sectioned coils with plunger, 256
Self-induction, effect of, 217
  in telegraph magnets, 214, 218
Siemens, differential coil-and-plunger, 250
  relay, 270
Siphon recorder, 270
Smith, plunger electromagnet, 253
Sparking, suppression of, 277
Steel, magnetization of, 58
  permeability of, 61
Stephenson, electric motors not practicable, 225
Stevens and Hardy, plunger electromagnet, 252
Stowletow, measurement of permeability, 59
Sturgeon, biographical sketch, 17
  experiments on bar magnets, 125
  experiments on leakage, 122

INDEX. 287

Sturgeon, first description of electromagnet, 7, 17
    magnetic circuit, 10
    polar extensions, 132
    polarized apparatus, 266
    portrait wanted, 279
    tubular cores, 158
Sturgeon's apparatus lost, 20
    first electromagnet, 18
    first experiments, 20
Surface magnetism, 108, 109

TIME-CONSTANT of electric circuit, 211, 213, 216, 218
Thomas, wire gauge table, 178
Thomson (Elihu), electromagnetic phenomena, 268
    (J J.), on effect of joints, 155
    (Sir Wm.), current meters, 255
    polarized magnets, 266
    range of action, 225
    rule for winding electromagnets, 183
    siphon recorder, 270
    winding galvanometer coils, 257
Thorpe's semaphore indicator, 275
Traction, formula for, 98
    in terms of weight of magnet, 98
Tractive power of magnets affected by surface contact, 135, 151
    integral formula for, 89
Trève, iron wire coil, 248
Tubular coils, action of, on a unit pole, 232
    attraction between, 243
    winding of, 256
Two magnetic fluids, doctrine of, 9
Tyndall, range of action, 225

VARLEY, copper sheath for magnet coils, 278
    electromagnet, 202
    iron-clad electromagnet, 188

Varley, polarized magnets, 266
Vaschy, coefficients of self-induction, 218
Vibrators, 274
Vincent, application of magnetic circuit in dynamo design, 12
Viscous hysteresis, 77
Von Feilitzsch, plungers of iron and steel, 244
    magnetism of long bars, 152
    measurement of permeability, 59
    tubular cores, 158
Von Kolke, distribution of magnetic lines, 137
Von Waltenhofen, attraction of two tubular coils, 243

WAGENER'S electromagnet, 204
    Wagner, vibrating mechanism, 274
Walmsley, magnetic reluctance of air, 148
Wheatstone, Henry's visit to, 38
    equalizer for telegraph instrument, 264
    oblique approach, 259
    polarized apparatus, 266
Winding a magnet in sections, 256
    calculation of, 95, 173, 183, 190
    coils in multiple arc, 258
    differential, 278
    effect of, on range of action, 248
    for constant pressure and for constant current, 182
    iron vs. copper wire, 202
    of tubular coils, 256
    position of coils, 193
    size of coils, 191
    thick versus thin wire, 78
    wire of graduated thickness, 257
Wire gauge and ampèreage table, 178
Wrought iron, magnetization of, 56, 64, 65

www.ingramcontent.com/pod-product-compliance
Lightning Source LLC
Chambersburg PA
CBHW031340230426
43670CB00006B/400